Mechanics of Flow Similarities

Claus Weiland

Mechanics of Flow Similarities

Springer

Claus Weiland
Bruckmühl, Germany
Head of Aerothermodynamics
Launcher Propulsion (retired)
EADS Space Transportation

ISBN 978-3-030-42929-4 ISBN 978-3-030-42930-0 (eBook)
https://doi.org/10.1007/978-3-030-42930-0

This Springer imprint is published by the registered company Springer Nature Switzerland AG
The registered company address is: Gewerbestrasse 11, 6330 Cham, Switzerland

To the woman,
I married 50 years ago.

Preface

Flows are physical fields where elements of material move under the influence of forces or physical interactions $\Rightarrow$ equations of motion. The field theories arose about the 1800s from the theory of potentials, for instance, the potentials of gravitational or electrical fields. The description of the field theory requires the elements of the infinitesimal calculus (integral and differential calculus). These were developed independently by Gottfried Wilhelm Leibniz (1646–1716) and Isaac Newton (1643–1727). Daniel Bernoulli (1700–1782) and Leonhard Euler (1707–1783), who were close friends, have worked and taught together in St. Petersburg around the year 1727. They were the first who have described flows and their field structure with the elements of the infinitesimal calculus. This has led to the system of partial differential equations, which characterizes inviscid flow fields, established by Leonhard Euler.

Between 1822 and 1845 these equations of motion were extended to viscous flows, today known as the Navier-Stokes equations. At that time comprehensive solutions either analytically or numerically were not possible. Therefore a large number of scientists and engineers had attempted to simplify the equations by sometimes far-reaching assumptions and/or the application of asymptotic as well as perturbation theories. Thereby solutions became possible for specific classes of problems. The results of the analysis of dimensions and the gasdynamic similarity, both are components of the mechanics of similarity, had accounted for such solutions.

The mechanics of similarity encompasses the analysis of dimensions, performed by various procedures, the gasdynamic similarity and the model technology. The analysis of dimensions delivers the dimensionless numbers[1] by which specific physical challenges could be described with a reduced number of variables. Thereby the assessment of the physical problems is facilitated. This book treats only fluid dynamic and thermodynamic issues. Up to the 60s of the last century, the contribution of the dimensionless numbers to the qualitative elucidation of several

[1]In the literature often the dimensionless numbers are also named similarity parameters. We use here only the term dimensionless numbers.

fluid dynamic and thermodynamic problems was substantial. Also for some particular problems quantitative solutions were obtained.

There are a lot of dimensionless numbers known, most of them are very specific and therefore have not a general significance. On the other hand from the analysis of dimensions regarding the Navier-Stokes equations, 21 dimensionless numbers can be extracted with an extended relevance, which characterize a lot of flows and various forms of the transfer of heat. These correlations bear the name of scientists or engineers, who first had derived and/or used these relations. For fluid dynamics and all sorts of heat transfer the discipline of the mechanics of similarity was so important that we highlight the historical background of all the persons who have contributed to the development of this discipline.

The classical gasdynamic similarity was developed mainly in the time period between the 1910s and the 1960s. The goal of this discipline was to find similarity rules which enables the aerodynamic engineer to perform transformations from existing flow fields to others, which meet geometrical and other specific flow field parameters. Most of these rules and findings do no longer play a role today, because modern powerful methods have superseded these tools. An exception of this is the Mach number independence principle of Oswatitsch, which still has a place in hypersonic flow considerations.

A recent investigation regarding the longitudinal aerodynamics of space vehicles has revealed that there exist other astonishing similarities for hypersonic and supersonic flow fields. It seems that most of the longitudinal aerodynamics obviously are independent from the geometrical configurations of the space vehicle considered, when simple transformations are applied. A section of this book is devoted to these new findings.

This book has been written for persons who are interested in the role the mechanics of similarity have played for the understanding of flows and heat transfer effects in the past and today. These could be graduate and doctoral students as well as teachers and lecturers. Further design and development engineers can find one or the other important insights.

Bruckmühl, Germany Claus Weiland
January 2020

Acknowledgements

When the idea is arising to write a monograph or a book one needs at least one person with which critical and constructive discussions about the size, volume and contents of such a work can be conducted. Further it is always a personal gain to get recommendations and suggestions on how to shape some sections and chapters. All that was done and provided by my colleague E. H. Hirschel, who additionally read all the chapters of the book for which I am much indebted to him.

Finally I wish to thank my wife for her support and patience.

Claus Weiland

Contents

Chapter 1
Introduction

Over a large number of decades, after Leonhard Euler (1707–1783) had formulated the system of equations for the description of convective processes in three-dimensional flow fields and further after its extension to flows with friction by Navier (1822), Poisson (1831), Saint-Vernant (1843) and Stokes (1845), a solution of such kind of systems of equations seemed to be impossible.

All efforts to attain analytical solutions were more or less futile. Even far-ranging simplifications and simplifying assumptions had only little success and did advance the exploitation of insights into the behavior of flow fields only in a lowly manner. At that time, essentially during the 19th century and the two first decades of the 20th century, questions were discussed, whether analytical methods could be developed, with which the number of parameters could be reduced, which describe complex flow fields. Further such methods should be able to transform the characteristics of a given flow field—for instance known by experimental investigations—to another flow field which satisfies geometrical similarity conditions.

The mechanics of similarity was the method with which such theoretical investigations could be conducted, [1–7].

1.1 The Environment of the Mechanics of Similarity

Today, we possess a lot of very potent methods for experimental investigations compared to those at earlier times. This holds for the wide field of diagnostics [8], the test facilities for all forms of flying vehicles (civil aircraft, combat aircraft, helicopter, space vehicle, etc.), ships and all sorts of propulsion systems (turbojets, ramjets, scramjets, gas turbines, rocket motors, etc.). Other powerful tools for the investigation of flow fields consists in discrete numerical methods approximating the related governing equations (Navier-Stokes equations, Euler equations, etc.), [9]. During the last four to five decades these methods have experienced a dramatic

C. Weiland, *Mechanics of Flow Similarities*,
https://doi.org/10.1007/978-3-030-42930-0_1

evolution. So, three-dimensional flow fields around geometrically ambitious configurations including various forms of thermodynamics (perfect gas, equilibrium real gas, non-equilibrium real gas) as well as advanced models for the simulation of turbulent flows,[1] can be calculated with much success, [9–11].

One of the main features of the mechanics of similarity consists in the ability to conclude from one well established flow field (e.g. by experimental testing) to another one, which has a similar geometry and satisfies all relevant dimensionless numbers.[2] However today, this task is performed relatively simple by discrete numerical methods, even if the computational costs (computer time and storage) are substantial. Anyway these decrease permanently due to the progress of the computer technology.

The physical understanding of a given flow field has at least two aspects:

- the first one is the quantitative insight into the details of the flow regarding the knowledge of all flow variables: velocity components, density, pressure and temperature given at every i^{th} coordinate point x_i, y_i, z_i.
- the second one consists in a qualitative comprehension of the main dependencies, which govern the formation of a given flow field.

There is no doubt, that the second item is supported by the mechanics of similarity and/or the analysis of dimensions, respectively. Presently, the significance of the mechanics of similarity lies in its application during the design phases of real projects, where various solutions and suggestions are discussed and their possible realization has to be estimated. This often will be conducted with so-called "design methods", which usually use some of the instruments of the mechanics of similarity. These methods give quick answers, are easy and cheap to handle, but their accuracy in many cases is limited. Nevertheless the design engineer gets an idea about the realization of the assumptions and proposals he did make. Most of the aforementioned issues have their relevance for fluid dynamic, aerodynamic as well as thermodynamic problems, whereas for the last item all aspects of heat exchange are relevant.

Generally the goal of the mechanics of similarity is as follows:

- to reduce the number of relevant parameters, which describe specific flow fields,
- to increase the state of knowledge about the dominating dependencies of flow fields,
- to transfer the controlling data of an established flow field to another one, which either has a fully similar geometry or a geometry given by an affine transformation.

To understand the environment of the mechanics of similarity some important statements can be formulated.

- There exists no compact theory, which enables one to determine all aspects of the mechanics of similarity.

[1]Of course, this does not mean, that the problem of turbulence is satisfactorily solved, but the applied models deliver solutions, which are more and more acceptable for practicable purposes.

[2]In the literature often the dimensionless numbers are also named similarity parameters. We use here only the term dimensionless numbers.

- Generally, the quality of the results of the mechanics of similarity depends strongly on the physics, which are employed for the description of a certain problem.
- Methods for the determination of dimensionless numbers are:
  - the analysis of dimensions of Buckingham, the Π theorem,
  - the fractional analysis method with the consideration of relations regarding energy, forces (momentum) and continuity,
  - the evaluation of the governing partial differential equations.

When working with the results of the mechanics of similarity the following rule has to be respected: For the use, formulation and interpretation of similarity laws and dimensionless numbers a certain carefulness is necessary, since the results are only meaningful, when the user has well understood the peculiarities of the respective flows.

1.2 The Contents of the Book

One of the main aspects of this book consists in the description of the methods by which the dimensionless numbers are derived, and further their physical interpretation. Another aspect highlights the gasdynamic similarity and its significance today. Most of the scientists and engineers, who have contributed to the development of the various parts of the mechanics of similarity are introduced.

After the Introduction the three subsequent chapters: the Buckingham's Π Theorem (Chap. 2), the Fractional Analysis Method (Chap. 3) and the Method of Differential Equations (Chap. 4), deal with the theory and analysis of dimensions in order to find characteristic quantities. In this sense the Π theorem is the most powerful method. In order to understand this theory some instructive examples are formulated in order to become more familiar with it. The fractional analysis method is a heuristic tool which was first addressed and developed by Lord Rayleigh at the end of the 19th century.

Of course to make the system of partial differential equations dimensionless delivers on the one hand a group of dimensionless numbers and when considering approximations of these equations on the other hand also reduced relations like the boundary layer equations (including boundary conditions) with possible similar solutions like the one of Blasius, [12], Chap. 4.

In Chap. 5 a classification of the dimensionless numbers is carried out with a view on the time period when they were formulated.

Chapter 6 lists the most interesting dimensionless numbers regarding fluid dynamics and thermodynamics. The historical background of the scientists and engineers is highlighted, whose names the dimensionless numbers bear. The significance of these parameters in the past and today is illustrated by several examples of applications.

Normally the engagement with the mechanics of similarity requires a fully geometrical similarity, which means that all the coordinates of a configuration are transformed by the same constant. The gasdynamic similarity deviates from this principle and considers only affine geometrical transformations, where every of the coordinates of a configuration are transformed by different values leading to stretched or warped geometries. Chapter 7 is dedicated to this topic.

The perturbation theory is used for the purpose of gasdynamic similarity to reduce the governing equations (- in order to make these easier to solve for a reduced regime of applications -) which permits also sometimes a linearization, as it is the case for subsonic and supersonic flows past thin or slender bodies at small angles of attack. For transonic and hypersonic flows a linearization can not be achieved. Oswatitsch's Mach number independence principle is an outcome of the gasdynamic similarity considerations for space vehicles in hypersonic flow, [13]. Recently, [14, 15], there was another similarity found regarding the lift, drag and pitching moment coefficients of space vehicles with different geometrical configurations. The details also are reported in Chap. 7.

One of the more significant applications of the mechanics of similarity is the model technique. In particular for the aerodynamic discipline this encompasses wind tunnel tests of airplanes and space planes as well as the aerodynamics of buildings, bridges, etc.. The same is true for the tests of engines and their components, like turbojets, ramjets, rocket motors, inlets, nozzles, etc.. In Chap. 8 the rules for model tests are formulated and an example of the wind tunnel tests regarding space vehicles is presented.

References

1. Sedov, L.J.: Similarity and Dimensional Methods in Mechanics. Academic, New York (1959)
2. Görtler, H.: Dimensionsanalyse, Theorie der physikalischen Dimensionen und Anwendungen. Springer, Berlin (1975)
3. Zierep, J.: Ähnlichkeitsgesetze und Modellregeln der Strömungsmechanik. G. Braun Verlag, Karlsruhe (1972)
4. Liepmann, H.W., Roshko, A.: Elements of Gasdynamics. Wiley, New York (1957)
5. Hayes, W.D., Probstein, R.F.: Hypersonic Flow Theory. Academic, New York (1966)
6. Hirschel, E.H.: Basics of Aerothermodynamics, 2nd edn. Springer, Berlin (2015)
7. Krause, E.: Strömungslehre, Gasdynamik und aerodynamisches Laboratorium. Teubner Verlag, Stuttgart (2003)
8. Dillmann, A.: Future Perspectives of Experimental Aerodynamics. DLR Göttingen. Paper presented at the celebration of E.H. Hirschel's 80th birthday, DLR Cologne (2014)
9. Hirschel, E.H., Krause, E. (eds.): 100 Volumes of 'Notes on Numerical Fluid Mechanics'. Notes on Numerical Fluid Mechanics and Multidisciplinary Design, vol. 100. Springer, Berlin (2009)
10. Rossow, C.-C.: Flugzeugentwurf in der Zukunft. DLR Braunschweig. Paper presented at the celebration of E.H. Hirschel's 80th birthday, DLR Cologne (2014)
11. Becker, K.: Numerical Simulation in Aerodynamics and Flight Physics. Airbus Bremen. Paper presented for celebrating E.H. Hirschel's 80th birthday, DLR Cologne (2014)
12. Blasius, H.: Grenzschichten in Flüssigkeiten mit kleiner Reibung. Z. Math. Phys. **56**, 1–37 (1908)
13. Oswatitsch, K.: ZAMP **2**, 249–264 (1951)
14. Weiland, C.: The aerodynamics of real space vehicles in the light of supersonic and hypersonic approximate theories. CEAS Space J. (2019). https://doi.org/10.1007/s12567-019-00264-w
15. Weiland, C.: Aerodynamic similarity of winged and lifting space vehicles. AIAA J. Aircraft. https://doi.org/10.2514/1.C035823 (2020). Accessed 30 Nov 2020

Chapter 2
Dimensional Analysis—Buckingham's Π Theorem

Edgar Buckingham was born on July 8, 1867 in Philadelphia, Pennsylvania, USA and died on April 25, 1940 in Washington, DC. He had his studies in Harvard, USA, at the University of Strasbourg, France and the University of Leipzig, Germany, where he received his PhD in 1893. From 1902 to 1906 he worked as a soil physicist at the US Bureau of Soils and afterwards accepted (1907) a post at the National Bureau of Standards, where he remained until his retirement in 1937. He performed a lot of work in the discipline of soil physics and published the results in some widely acknowledged papers, [1, 2]. In 1914 he presented the paper [3], in which he summed up the theory of the Π theorem with its fundamental significance for the analysis of dimensions, see also [4].

Edgar Buckingham

2.1 The Π Theorem of Buckingham

Let us consider a number of n physical variables, which have dimensions, $K_1, K_2, \ldots, K_n$, where the relation exists

$$f(K_1, K_2, \ldots K_n) = 0\,. \tag{2.1}$$

Further we define $m \leq n$ variables with dimensions $L_1, L_2, \ldots L_m$, which we denote basic dimensions with the property that none of these variables can be constituted through power products of the remaining others. Each of the dimensions of

C. Weiland, *Mechanics of Flow Similarities*,
https://doi.org/10.1007/978-3-030-42930-0_2

the variables $[K_i]$ can be presented by power products of the variables L_j, see [3, 5, 6].

Therefore we have

$$\begin{aligned}
[K_1] &= L_1^{a_{11}} \ L_2^{a_{21}} \ \cdots \ L_m^{a_{m1}} , \\
[K_2] &= L_1^{a_{12}} \ L_2^{a_{22}} \ \cdots \ L_m^{a_{m2}} , \\
\vdots \quad & \quad \vdots \qquad \vdots \qquad \ddots \quad \vdots \\
[K_n] &= L_1^{a_{1n}} \ L_2^{a_{2n}} \ \cdots \ L_m^{a_{mn}} .
\end{aligned} \tag{2.2}$$

Now, the question arises, if there exist dimensionless forms of the kind

$$\Pi = K_1^{b_1} \cdot K_2^{b_2} \cdots K_n^{b_n} \tag{2.3}$$

and how much.

Considering the dimensions in Eq. (2.3) and combining these with Eqs. (2.2) we obtain

$$\begin{aligned}
[\Pi] &= \left(L_1^{a_{11}} \cdot L_2^{a_{21}} \cdots L_m^{a_{m1}}\right)^{b_1} \cdots \left(L_1^{a_{1n}} \cdot L_2^{a_{2n}} \cdots L_m^{a_{mn}}\right)^{b_n} , \\
&= L_1^0 \cdot L_2^0 \cdots L_m^0 ,
\end{aligned} \tag{2.4}$$

where the exponent 0 in the basic dimensions stands for the generation of the dimensionless numbers.

Comparing the exponents of the basic dimensions (L_i) Eq. (2.4) yields

$$\begin{aligned}
a_{11}b_1 + a_{12}b_2 + \cdots + a_{1n}b_n &= 0 , \\
a_{21}b_1 + a_{22}b_2 + \cdots + a_{2n}b_n &= 0 , \\
\vdots \qquad \vdots \qquad \ddots \qquad & \ \vdots \\
a_{m1}b_1 + a_{m2}b_2 + \cdots + a_{mn}b_n &= 0 .
\end{aligned} \tag{2.5}$$

This is a homogeneous linear algebraic system for m equations with n unknowns $b_1, b_2, \cdots b_n$.

$$\mathbf{A} \cdot \mathbf{b} = \mathbf{0} . \tag{2.6}$$

The coefficient matrix $\mathbf{A}$ of this linear algebraic system has the form

$$\mathbf{A} = \begin{Bmatrix}
a_{11} & a_{12} & \cdots & a_{1n} \\
a_{21} & a_{22} & \cdots & a_{2n} \\
\vdots & \vdots & \ddots & \vdots \\
a_{m1} & a_{m2} & \cdots & a_{mn}
\end{Bmatrix} \tag{2.7}$$

From the linear algebra discipline we know that the number of independent linear solutions p depend on the rank r of the matrix Eq. (2.7), where the rank of a matrix is given by the order of the highest non-vanishing determinant. So we find $p = n - r$.

Now, Buckingham's Π theorem[1] says:

Given are n dimensional variables $K_1, K_2, \ldots, K_n$ and a correlation between them. Then, there exist exactly $p = n - r$ dimensionless numbers Π_i.

The relation

$$f(\Pi_1, \Pi_2, \ldots, \Pi_p) = 0 \tag{2.8}$$

solves the problem.

2.2 Illustrative Examples

In this section some selected examples are presented, where Buckingham's Π theorem is applied in order to reduce the number of independent variables and to increase the physical understanding of the flow considered. In general, in order to attain the most promising result, one should try to define the physical problem such,[2] that the number of dimensionless numbers p is as low as possible.[3] First of all it is necessary to define a set of basic dimensions. In mechanics a natural one consists of the mass m_a, the length l and the time t with the dimensions kg, m, s. But it can be useful in order to achieve optimal results when applying the Π theorem to turn to other, more appropriate systems. This could be, for example, a set with the length l, the time t and the force f with the dimensions m, s, $N = \text{kg m/s}^2$, or with the length l, the velocity u and the mass m_a with the dimensions m, m/s, kg.

Below we present at first four mechanical problems (basis: momentum equation), where the three basic dimensions length l, time t and mass m_a are used. Then a combined thermodynamic/mechanical problem (basis: energy equation) with the four basic dimensions[4] l, t, m_a, T is examined. The procedures for the treatment of the examples "A Generalized Flow Problem" and "Heat Transfer over a Heated Flat Plate", given below are executed in much detail. This shall ensure that the reader obtains an enhanced understanding of the method and that he becomes somewhat familiar with it. The remaining examples have interesting flow physical backgrounds.

---

[1] Görtler in [6] questioned that the Π theorem was formulated first by E. Buckingham. He gives in this book a historical survey about the origination of the Π theorem.

[2] The number of physical variables n should be restricted to the most important ones.

[3] $p = 0$ would be optimal, because no dimensionless number exists and the physical problem can be directly solved up to a constant by the Π theorem.

[4] The three basic dimensions are extended by the temperature T.

Table 2.1 The matrix of dimensions for a general flow problem

	l	t	v	b	ρ	μ	Δp
kg	0	0	0	0	1	1	1
m	1	0	1	1	−3	−1	−1
s	0	1	−1	−2	0	−1	−2

The results of Buckingham's method will be presented there only in a standard manner.

Finally we deal with a problem where $p = 1$, and another problem where $p = 0$ and the set of basic dimensions is given by f, l, t.

2.2.1 A Generalized Flow Problem

The generalized flow problem is defined by n = 7 characteristic variables, length l, time t, velocity v, acceleration b, difference of pressure Δp, density ρ and dynamic viscosity μ. The basic dimensions for this kind of mechanical problem are length l, time t, mass m_a with the dimensions $[l] \doteqdot$ m, $[t] \doteqdot$ s, $[m_a] \doteqdot$ kg.

The rank of the dimensional matrix, Table 2.1, is $r = 3$. Therefore $p = n - r = 4$ dimensionless numbers Π_i exist. When we have a look at the matrix of dimensions we realize that we have seven unknown coefficients $k_1, \ldots, k_7$ and only three equations. This gives us some freedom for the determination of four of the coefficients k_i. These coefficients should be chosen in view of the physical problem to be considered, which means that some kind of physical intuition or inspiration must be available.[5] A promising procedure consists in the definition of a set of basic quantities, driven by physical experience, which are specific for each dimensionless numbers Π_i. We will demonstrate that in the frame of the calculation procedure below.

(a) For the first dimensionless number Π_1 we choose the set of basic quantities l, t, v, ρ and we find the linear algebraic system derived from the matrix of dimensions (Table 2.1) to

$$\begin{aligned} 0 \cdot k_1 + 0 \cdot k_2 + 0 \cdot k_3 + 1 \cdot k_5 &= 0\,, \\ 1 \cdot k_1 + 0 \cdot k_2 + 1 \cdot k_3 - 3 \cdot k_5 &= 0\,, \\ 0 \cdot k_1 + 1 \cdot k_2 - 1 \cdot k_3 + 0 \cdot k_5 &= 0\,. \end{aligned} \tag{2.9}$$

With the assumed value $k_1 = 1$ the evaluation of the system Eqs. (2.9) yields $k_2 = -1$, $k_3 = -1$, $k_5 = 0$ and

$$\Pi_1 = l^1 t^{-1} v^{-1} \rho^0 = \frac{l}{vt} = Str, \qquad \text{Strouhal number.} \tag{2.10}$$

[5]Other authors notice that some coefficients have to be guessed.

(b) For the second dimensionless number Π_2 a reasonable set of basic quantities is l, v, ρ, μ. The linear algebra system reads then

$$\begin{aligned} 0 \cdot k_1 + 0 \cdot k_3 + 1 \cdot k_5 + 1 \cdot k_6 &= 0\,, \\ 1 \cdot k_1 + 1 \cdot k_3 - 3 \cdot k_5 - 1 \cdot k_6 &= 0\,, \\ 0 \cdot k_1 - 1 \cdot k_3 + 0 \cdot k_5 - 1 \cdot k_6 &= 0\,. \end{aligned} \tag{2.11}$$

Assuming again that $k_1 = 1$ we find $k_3 = 1, k_5 = 1, k_6 = -1$ and

$$\Pi_2 = l^1 v^1 \rho^1 \mu^{-1} = \frac{\rho v l}{\mu} = Re, \qquad \text{Reynolds number.} \tag{2.12}$$

(c) For the third dimensionless number Π_3 a selection of the set of basic quantities like l, v, ρ, Δp seems to be obvious. The linear algebra system derived from Table 2.1 has the form

$$\begin{aligned} 0 \cdot k_1 + 0 \cdot k_3 + 1 \cdot k_5 + 1 \cdot k_7 &= 0\,, \\ 1 \cdot k_1 + 1 \cdot k_3 - 3 \cdot k_5 - 1 \cdot k_7 &= 0\,, \\ 0 \cdot k_1 - 1 \cdot k_3 + 0 \cdot k_5 - 2 \cdot k_7 &= 0\,. \end{aligned} \tag{2.13}$$

In that case it should be appropriate to assume $k_3 = -2$. With that it follows for the remaining three coefficients by evaluation of Eqs. (2.13) $k_1 = 0, k_5 = -1, k_7 = 1$ and

$$\Pi_3 = l^0 v^{-2} \rho^{-1} \Delta p^1 = \frac{\Delta p}{\rho v^2} = Eu, \qquad \text{Euler number.} \tag{2.14}$$

(d) Finally, the fourth dimensionless number Π_4 can be determined with the help of the following set of basic quantities: l, v, b, ρ. The linear algebra system derived from Table 2.1 now has the form

$$\begin{aligned} 0 \cdot k_1 + 0 \cdot k_3 + 0 \cdot k_4 + 1 \cdot k_5 &= 0\,, \\ 1 \cdot k_1 + 1 \cdot k_3 + 1 \cdot k_4 - 3 \cdot k_5 &= 0\,, \\ 0 \cdot k_1 - 1 \cdot k_3 - 2 \cdot k_4 + 0 \cdot k_5 &= 0\,. \end{aligned} \tag{2.15}$$

Choosing $k_1 = -1$ it follows $k_3 = 2, k_4 = -1, k_5 = 0$ and

$$\Pi_4 = l^{-1} v^2 g{-}1 \rho^0 \Delta p^1 = \frac{v^2}{lg} = Fr, \qquad \text{Froude number.} \tag{2.16}$$

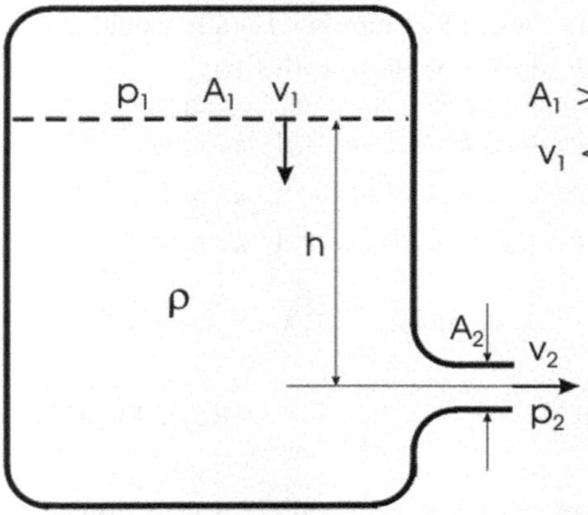

Fig. 2.1 Sketch of the outflow of a tank

2.2.2 *Outflow of a Tank*

The flow situation is shown in Fig. 2.1, [5, 7]. This problem is governed by the six characteristic variables ($n = 6$): velocity v, density ρ, gravity g, difference of pressure $\Delta p = p_1 - p_2$, the area at the outflow orifice A, and the altitude h.

$$f(v, \rho, g, \Delta p, A, h) = 0 \tag{2.17}$$

The evaluation procedure is as in the example above. The rank of the matrix of dimensions, Table 2.2, is $r = 3$ and therefore we have $p = n - r = 6 - 3 = 3$ dimensionless numbers Π_i.

(a) For the first dimensionless number Π_1 we define the set of basic quantities to be v, ρ, g, Δp. In the sense of the approach of the example above (Sect. 2.2.1) we make the guess $k_1 = -2$ and find by evaluation of the correspondent linear algebraic system $k_2 = -1$, $k_3 = 0$, $k_4 = 1$. Then we obtain

$$\Pi_1 = v^{-2}\rho^{-1}g^0\Delta p^1 = \frac{\Delta p}{\rho v^2} = Eu \;, \qquad \text{Euler number.} \tag{2.18}$$

(b) The set of basic quantities for the determination of the second dimensionless number Π_2 can be defined by v, ρ, g, h. With the assumed value for the first coefficient

Table 2.2 The matrix of dimensions for an outflow of a tank

	v	ρ	g	Δp	A	h
kg	0	1	0	1	0	0
m	1	−3	1	−1	2	1
s	−1	0	−2	−2	0	0

$k_1 = 1$ we obtain $k_2 = 0$, $k_3 = -1/2$, $k_4 = -1/2$ and[6]

$$\Pi_2 = v^1 \rho^0 g^{-1/2} h^{-1/2} = \frac{v}{\sqrt{gh}} = Fr' , \qquad \text{Froude number} . \tag{2.19}$$

If we assume $\Pi_2 = \sqrt{2}$, we find Toricelli's formula

$$v = \sqrt{2gh} . \tag{2.20}$$

(c) We use for the determination of the third dimensionless number Π_3 the set of basic quantities v, g, A, h. With the physical guess for the first coefficient $k_1 = 1$ we find $k_3 = -1/2$, $k_5 = 1$, $k_6 = -5/2$ and

$$\Pi_3 = v^1 g^{-1/2} A^1 h^{-5/2} = \frac{v A}{\sqrt{g}\, h^{5/2}} . \tag{2.21}$$

To understand the nature of the flow we have reduced the number of variables from 6 to 3 by the above operations. For a given flow situation the dimensionless numbers Π_i are constants and therefore also any correlation of them. If we make the assumption[7]

$$\begin{aligned} C_1 &= f_1\left(\frac{v}{\sqrt{gh}}\right) + g_1\left(\frac{\Delta p}{\rho\, v^2}\right) = \\ &= 2(\Pi_2^{-2} + \Pi_1^1) , \end{aligned}$$

then we obtain

$$\sqrt{C_1}\; v = \sqrt{2gh + \frac{2\Delta p}{\rho}} . \tag{2.22}$$

Equation (2.22) is the well-known outflow formula of a pressurized tank. Another correlation yields

$$\begin{aligned} C_2 &= f_2\left(\frac{v}{\sqrt{gh}}\right) \cdot g_2\left(\frac{vA}{\sqrt{g}\, h^{5/2}}\right) \\ &= \Pi_2^{-2} \cdot \Pi_3^1, \\ C_2\, v &= \frac{A\sqrt{g}}{h^{3/2}} , \end{aligned} \tag{2.23}$$

[6]There are two definitions of the Froude number: $Fr = v^2/gh$ and $Fr' = v/\sqrt{gh}$ depending on the application case, see Sect. 6.2.

[7]Of course we did it in view of the result we know.

and with $\dot{m} = \rho v A$ we obtain

$$C_2 \, \dot{m} = \rho \sqrt{2gh} \cdot A \left(\frac{A}{h^2 \sqrt{2}} \right) . \tag{2.24}$$

If C_2 is given by $C_2 = \dfrac{A}{h^2 \sqrt{2}}$, we have Toricelli's formula. C_1 and C_2 are constants, see Eqs. (2.22)–(2.24).

This is a very impressive and specific example, which demonstrates the power of the analysis of dimensions, which is part of the mechanics of flow similarity. But in general this method does not provide mathematical relations of functional dependencies, instead it reduces the number of characteristic variables.

2.2.3 *Pipe Flow*

The characteristic variables ($n = 7$) for a pipe flow are the average velocity v_a, the density ρ, the difference of pressure Δp, the dynamic viscosity μ, the pipe length l, the wall roughness k and the pipe diameter D, Fig. 2.2.

$$f(v_a, \rho, \Delta p, \mu, l, k, D) = 0 \tag{2.25}$$

The rank of the matrix of dimensions, Table 2.3, again is $r = 3$ and therefore we have $p = n - r = 7 - 3 = 4$ dimensionless numbers Π_i.

(a) For the first dimensionless number Π_1 the set of basic quantities is given by $v_a, \rho, \Delta p, D$. With the guess $k_1 = -2$ we find by evaluation of the correspondent linear algebraic system $k_2 = -1$, $k_3 = 1$, $k_7 = 0$ and

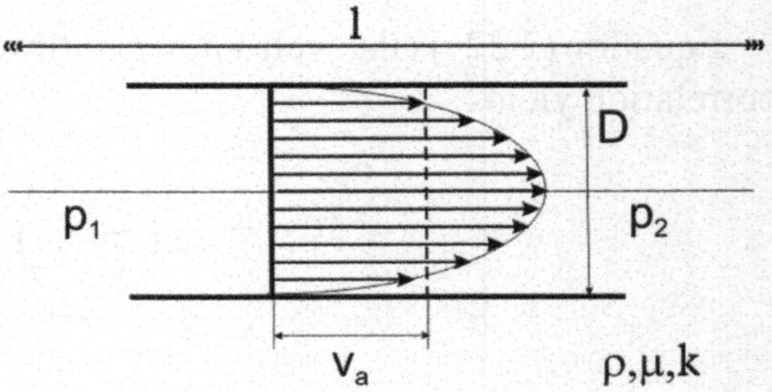

Fig. 2.2 Sketch of a fully developed laminar pipe flow

Table 2.3 The matrix of dimensions for a pipe flow

	v_a	ρ	Δp	μ	l	k	D
kg	0	1	1	1	0	0	0
m	1	−3	−1	−1	1	1	1
s	−1	0	−2	−1	0	0	0

$$\Pi_1 = v_a^{-2} \rho^{-1} \Delta p^1 D^0 = \frac{\Delta p}{\rho v_a^2} = Eu \ . \tag{2.26}$$

(b) The used set of basic quantities for the dimensionless number Π_2 is v_a, ρ, μ, D. With $k_1 = 1$ one obtains $k_2 = 1, k_4 = -1, k_7 = 1$ and

$$\Pi_2 = v_a^1 \rho^1 \mu^{-1} D^1 = \frac{\rho v_a D}{\mu} = Re \ . \tag{2.27}$$

(c) In this case the used set of basic quantities for the dimensionless number Π_3 is v_a, ρ, l, D. With $k_5 = 1$ one has $k_1 = 0, k_2 = 0, k_7 = -1$, hence

$$\Pi_3 = v_a^0 \rho^0 l^1 D^{-1} = \frac{l}{D} \ . \tag{2.28}$$

(d) Finally for the dimensionless number Π_4 we use correspondingly to case **(c)** the set of basic quantities v_a, ρ, k, D. With $k_6 = 1$ it follows $k_1 = 0, k_2 = 0, k_7 = -1$ and

$$\Pi_4 = v_a^0 \rho^0 k^1 D^{-1} = \frac{k}{D} \ . \tag{2.29}$$

The pressure loss in the pipe can then be described by the three variables

$$\frac{\Delta p}{\rho v_a^2} = f\left(Re, \frac{l}{D}, \frac{k}{D}\right) \ . \tag{2.30}$$

2.2.4 *Fluid Dynamical Drag of a Ship*

In this example the drag force D of a ship is investigated. The characteristic variables ($n = 7$) of this flow are the velocity v, the density ρ, the gravity g, the dynamic viscosity μ, the length of the ship L, the breadth of the ship B and the draft of the ship t, Fig. 2.3.

$$D = f(v, \rho, g, \mu, L, B, t) \tag{2.31}$$

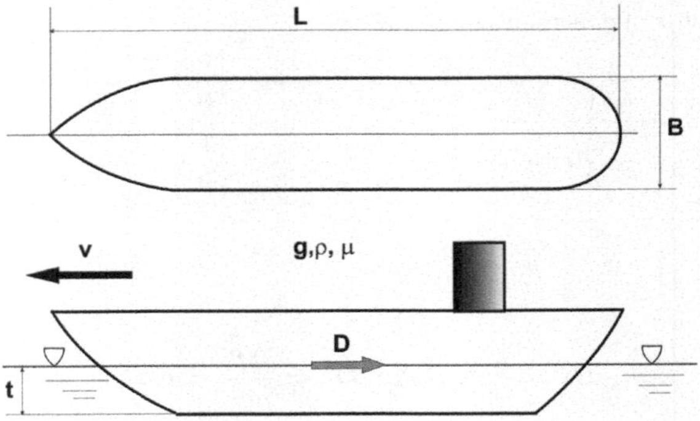

Fig. 2.3 Sketch of a ship movement

Table 2.4 The matrix of dimensions for the fluid dynamical drag of a ship

	v	ρ	g	μ	L	B	t
kg	0	1	0	1	0	0	0
m	1	−3	1	−1	1	1	1
s	−1	0	−2	−1	0	0	0

The evaluation procedure of the matrix of dimensions, Table 2.4, is the same as in the examples before. Therefore we list here only the results.

(a) For Π_1 the set of basic quantities is v, ρ, g, B. The related coefficients are $k_1 = 2, k_2 = 0, k_3 = -1, k_6 = -1$ and

$$\Pi_1 = v^2\rho^0 g^{-1} B^{-1} = \frac{v^2}{gB} = Fr\ . \tag{2.32}$$

(b) For Π_2 the set of basic quantities is v, ρ, μ, B. The related coefficients are $k_1 = 1, k_2 = 1, k_4 = -1, k_6 = 1$ and

$$\Pi_2 = v^1\rho^1\mu^{-1}B^1 = \frac{\rho v B}{\mu} = Re\ . \tag{2.33}$$

(c) For Π_3 the set of basic quantities is v, ρ, L, B. The related coefficients are $k_1 = 0, k_2 = 0, k_5 = 1, k_6 = -1$ and

$$\Pi_3 = v^0\rho^0 L^1 B^{-1} = \frac{L}{B}\ . \tag{2.34}$$

(d) For Π_4 the set of basic quantities is v, ρ, B, t. The related coefficients are $k_1 = 0, k_2 = 0, k_6 = -1, k_7 = 1$ and

$$\Pi_4 = v^0 \rho^0 B^{-1} t^1 = \frac{t}{B} . \tag{2.35}$$

The function f, Eq. (2.31), transfers to the dimensionless function g, which is now a function of only four variables, and for the dimensionless drag force we obtain

$$\frac{D}{\rho g L B t} = g\left(Fr, Re, \frac{L}{B}, \frac{t}{B}\right) . \tag{2.36}$$

Full model similarity is given, if the magnitude of all the dimensionless numbers are the same for the model and the prototype flow. It seems to be simple to obtain the similarity with the geometrical parameters L/B and t/B for the ship model. A constant Reynolds number requires for the model flow an increase of the model velocity v by the same value as the model is geometrically scaled down. But in the Froude number the velocity appears quadratically, which means that the Reynolds number and the Froude number can not be met simultaneously by such operations.

2.2.5 Heat Transfer Over a Heated Flat Plate

This heat transfer problem is characterized by $n = 12$ characteristic variables, α the heat transfer coefficient, ΔT the temperature difference between the wall of the plate and the freestream $\Delta T = T_w - T_\infty$, T_∞ the freestream temperature, c_p, c_v the specific heats, λ the thermal conductivity, v_∞ the freestream velocity, ρ_∞ the freestream density, μ the dynamic viscosity, g the gravitational acceleration, L, B the length and the width of the plate, Fig. 2.4, [5]. The function f describes the problem

$$f(\alpha, \Delta T, T_\infty, c_p, c_v, \lambda, v_\infty, \rho_\infty, \mu, g, L, B) = 0 \tag{2.37}$$

Four basic dimensions are now available, the length l, the time t, the mass m_a and the temperature T, ($[l] \doteqdot$ m, $[t] \doteqdot$ s, $[m_a] \doteqdot$ kg, $[T] \doteqdot$ K). Table 2.5 shows the matrix of dimensions. The rank of this matrix is $r = 4$ and therefore the number of dimensionless numbers Π_i is $p = n - r = 12 - 4 = 8$. The following procedure is conducted in the same way as outlined in Sect. 2.2.1. We obtain

(a) The set of basic quantities for the first dimensionless number Π_1 is chosen by α, ΔT, λ, ρ_∞, L. With that the linear algebra system derived from Table 2.5 has the form

Table 2.5 The matrix of dimensions for the heat transfer problem

	α	ΔT	T_∞	c_p	c_v	λ	v_∞	ρ_∞	μ	g	L	B
kg	1	0	0	0	0	1	0	1	1	0	0	0
m	0	0	0	2	2	1	1	−3	−1	1	1	1
s	−3	0	0	−2	−2	−3	−1	0	−1	−2	0	0
K	−1	1	1	−1	−1	−1	0	0	0	0	0	0

$$\begin{aligned}
1 \cdot k_1 + 0 \cdot k_2 + 1 \cdot k_6 + 1 \cdot k_8 + 0 \cdot k_{11} &= 0\,,\\
0 \cdot k_1 + 0 \cdot k_2 + 1 \cdot k_6 - 3 \cdot k_8 + 1 \cdot k_{11} &= 0\,,\\
-3 \cdot k_1 + 0 \cdot k_2 - 3 \cdot k_6 + 0 \cdot k_8 + 0 \cdot k_{11} &= 0\,,\\
-1 \cdot k_1 + 1 \cdot k_2 - 1 \cdot k_6 + 0 \cdot k_8 + 0 \cdot k_{11} &= 0\,. \qquad (2.38)
\end{aligned}$$

When we choose $k_1 = 1$ we find by evaluation of this linear system $k_2 = 0$, $k_6 = -1$, $k_8 = 0$ and $k_{11} = 1$ which leads to

$$\Pi_1 = \alpha^1\, \Delta T^0\, \lambda^{-1}\, \rho_\infty^0\, L^1 = \frac{\alpha\, L}{\lambda} = Nu\,, \qquad \text{Nusselt number}\,. \qquad (2.39)$$

The Nusselt number compares the heat transport due to convection to a rigid wall with the heat transport due to conduction in the fluid. Or the other way around, the Nusselt number indicates how much the heat transport due to convection is larger than the heat transport due to conduction in the fluid, see Sect. 6.9.

(b) For Π_2 the set of basic quantities is chosen to ΔT, c_p, v_∞, ρ_∞, L and we obtain for the system of linear equations

$$\begin{aligned}
0 \cdot k_2 + 0 \cdot k_4 + 0 \cdot k_7 + 1 \cdot k_8 + 0 \cdot k_{11} &= 0\,,\\
0 \cdot k_2 + 2 \cdot k_4 + 1 \cdot k_7 - 3 \cdot k_8 + 1 \cdot k_{11} &= 0\,,\\
0 \cdot k_2 - 2 \cdot k_4 - 1 \cdot k_7 + 0 \cdot k_8 + 0 \cdot k_{11} &= 0\,,\\
1 \cdot k_2 - 1 \cdot k_4 + 0 \cdot k_7 + 0 \cdot k_8 + 0 \cdot k_{11} &= 0\,. \qquad (2.40)
\end{aligned}$$

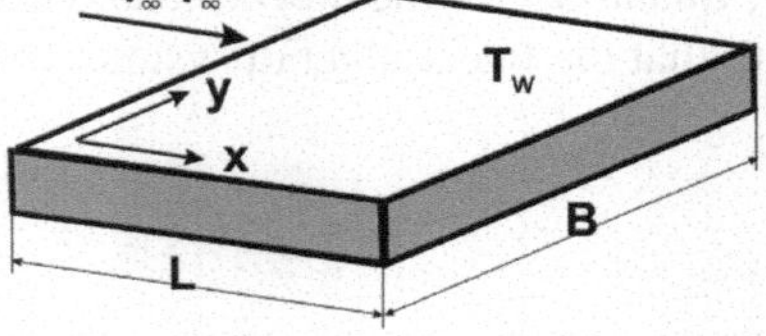

Fig. 2.4 Flow over a heated flat plate

Assuming that $k_2 = -1$ we get from the resolution of Eqs. (2.40) $k_4 = -1, k_7 = 2$, $k_8 = 0$, $k_{11} = 0$ and

$$\Pi_2 = \Delta T^{-1}\, c_p^{-1}\, v_\infty^2\, \rho_\infty^0\, L^0 = \frac{v_\infty^2}{c_p\, \Delta T} = \tilde{Ec}\,, \qquad \text{Eckert number}\,. \tag{2.41}$$

In Sect. 6.7 we have defined the Eckert number also with T instead of ΔT, $Ec = \frac{v_\infty^2}{c_p\, T}$, which then for perfect gas is connected with the Mach number by $Ec = (\gamma - 1)M^2$, [8].

(c) The next dimensionless number Π_3 is calculated by applying the basic quantities ΔT, T_∞, ρ_∞, μ, g, L. The set of linear equations has the form

$$\begin{aligned} 0 \cdot k_2 + 0 \cdot k_3 + 1 \cdot k_8 + 1 \cdot k_9 + 0 \cdot k_{10} + 0 \cdot k_{11} &= 0\,,\\ 0 \cdot k_2 + 0 \cdot k_3 - 3 \cdot k_8 - 1 \cdot k_9 + 1 \cdot k_{10} + 1 \cdot k_{11} &= 0\,,\\ 0 \cdot k_2 + 0 \cdot k_3 + 0 \cdot k_8 - 1 \cdot k_9 - 2 \cdot k_{10} + 0 \cdot k_{11} &= 0\,,\\ 1 \cdot k_2 + 1 \cdot k_3 + 0 \cdot k_O + 0 \cdot k_9 + 0 \cdot k_{10} + 0 \cdot k_{11} &= 0\,. \end{aligned} \tag{2.42}$$

In that case we have the freedom to pick two of the 6 coefficients used in Eqs. (2.42), for example $k_2 = 1$ and $k_8 = 2$ and find for the remaining ones $k_3 = -1$, $k_9 = -2$, $k_{10} = 1$, $k_{11} = 3$ and[8]

$$\Pi_3 = \Delta T^1\, T_\infty^{-1}\, \rho_\infty^2\, \mu^{-2}\, g^1\, L^3 = \frac{g\, l^3\, \Delta T}{\nu^2\, T_\infty} = Gr\,, \qquad \text{Grashof number}\,. \tag{2.43}$$

(d) In order to determine Π_4 we may choose the basic quantities v_∞, ρ_∞, g. L, and get

$$\begin{aligned} 0 \cdot k_7 + 1 \cdot k_8 + 0 \cdot k_{10} + 0 \cdot k_{11} &= 0\,,\\ 1 \cdot k_7 - 3 \cdot k_8 + 1 \cdot k_{10} + 1 \cdot k_{11} &= 0\,,\\ -1 \cdot k_7 + 0 \cdot k_8 - 2 \cdot k_{10} + 0 \cdot k_{11} &= 0\,,\\ 0 \cdot k_7 + 0 \cdot k_8 + 0 \cdot k_{10} + 0 \cdot k_{11} &= 0\,. \end{aligned} \tag{2.44}$$

If we choose $k_{11} = -1$ we obtain for the other involved coefficients $k_7 = 2$, $k_8 = 0$, $K_{10} = -1$ leading[9] to

---

[8] This definition is true, if the coefficient of thermal expansion β is chosen to be $\beta = 1/T_\infty$, which is a good approximation for the perfect gas case, see Sect. 6.4.

[9] For ship design and ship flow often the square root of the Froude number is used, $Fr' = \sqrt{Fr} = v_\infty/\sqrt{g\, L}$, see Sect. 6.2.

$$\Pi_4 = v_\infty^2 \, \rho_\infty^0 \, g^{-1} \, L^{-1} = \frac{v_\infty^2}{g\,L} = Fr \; . \qquad \text{Froude number} \; . \tag{2.45}$$

The remaining four dimensional parameters $\Pi_5 - \Pi_8$ are treated only in short.

(e) The basic quantities are c_p, λ, ρ_∞, μ. The coefficients k_4, k_6, k_8, k_9 have to be determined. The resolution of the related system of linear equations yields: $k_4 = 1$, $k_6 = -1$, $k_8 = 0$, $k_9 = 1$. The dimensionless number Π_5 is

$$\Pi_5 = c_p^1 \, \lambda^{-1} \, \rho_\infty^0 \, \mu^1 = \frac{c_p \, \mu}{\lambda} = \frac{\nu}{k} = Pr \; , \qquad \text{Prandtl number} \; , \tag{2.46}$$

with $\nu = \mu/\rho_\infty$ the kinematic viscosity and $k = \lambda/\rho_\infty \, c_p$ the thermal diffusivity. The Prandtl number denotes the ratio between the transport coefficients of momentum and energy.

(f) The basic quantities are v_∞, ρ_∞, μ, L, B. The coefficients $k_7, k_8, k_9, k_{11}, k_{12}$ are relevant for the problem. The corresponding set of linear equations provides the solution $k_7 = 1$, $k_8 = 1$, $k_9 = -1$, $k_{11} = 1$, $k_{12} = 0$ and

$$\Pi_6 = v_\infty^1 \, \rho_\infty^1 \, \mu^{-1} \, L^1 \, B^0 = \frac{v_\infty \, \rho_\infty \, L}{\mu} = Re \; , \qquad \text{Reynolds number} \; . \tag{2.47}$$

(g) The basic quantities are c_p, c_v, v_∞, ρ_∞. The corresponding coefficients are k_4, k_5, k_7, k_8. This leads with the well-known procedure to $k_4 = 1$, $k_5 = -1$, $k_7 = 0$, $k_8 = 0$ and

$$\Pi_7 = c_p^1 \, c_v^{-1} v_\infty^0 \, \rho_\infty^0 = \frac{c_p}{c_v} = \gamma \; , \qquad \text{ratio of specific heats} \; . \tag{2.48}$$

(h) The determination of the final dimensionless number Π_8 can be attained with the basic quantities v_∞, ρ_∞, L, B, where the corresponding coefficients are then k_7, k_8, k_{11}, k_{12}. This results in $k_7 = 0$, $k_8 = 0$, $k_{11} = 1$, $k_{12} = -1$ and

$$\Pi_8 = v_\infty^0 \, \rho_\infty^0 \, L^1 \, B^{-1} = \frac{L}{B} \; . \tag{2.49}$$

With the 8 dimensionless numbers Π_i the 12 arguments of the function f of Eq. (2.37) can be reduced to

$$\begin{aligned} 0 &= f_1(\Pi_1 \cdots \Pi_8) \qquad \text{or} \\ Nu &= f_2(\Pi_2 \cdots \Pi_8) = \\ &= f_2\left(Ec, Gr, Fr, Pr, Re, \gamma, \frac{L}{B}\right) . \end{aligned} \tag{2.50}$$

2.2.6 Laminar, Incompressible Boundary Layer Flow Along a Flat Plate (Blasius Boundary Layer)

From boundary layer theory we know that the thickness δ of a laminar, incompressible boundary layer along a flat plate (Blasius boundary layer) depends on the x-coordinate, the freestream velocity v_∞ and the kinematic viscosity ν. Since the pressure p in this boundary layer is constant throughout the flow field, there also is no dependency of δ on the density ρ, [8, 9]. Therefore we have, [6]

$$\delta = f(x, v_\infty, \nu) . \tag{2.51}$$

It is obvious that the rank of the matrix of dimensions, Table 2.6, is $r = 2$ and with $n = 3$ one obtains $p = n - r = 1$. This means that there exists one dimensionless number Π_1. For solving the problem one has, due to the Π theorem, to look for a power product of the dimension l of δ, which is in this case directly given by x and it follows

$$\delta = xg(\Pi_1) . \tag{2.52}$$

Further, when evaluating the matrix of dimensions, Table 2.6, the dimensionless power product with $k_1 = 1,\ k_2 = 1,\ k_3 = -1$ is given by $\Pi_1 = \dfrac{x v_\infty}{\nu}$ and one obtains

$$\delta = x\, g\left(\frac{x v_\infty}{\nu}\right) . \tag{2.53}$$

If one remembers that the boundary layer theory says that δ is proportional to $\sqrt{\nu}$, [9], one finds

$$\delta = const \cdot x \sqrt{\frac{\nu}{x v_\infty}} = const \cdot \sqrt{\frac{x\nu}{v_\infty}} \quad \text{or} \tag{2.54}$$

$$\frac{\delta}{x} \sim \frac{1}{\sqrt{Re_x}} , \tag{2.55}$$

where Re_x is the local Reynolds number formulated with the x-coordinate, [8].

Table 2.6 The matrix of dimensions for the Blasius boundary layer

	x	v_∞	ν	δ
kg	0	0	0	0
m	1	1	2	1
s	0	−1	−1	0

Table 2.7 Force in the wire of a rotating mass

	l_w	$m_{a,w}$	v_w	f_w
kg m/s^2	0	1	0	1
m	1	−1	1	0
s	0	2	−1	0

2.2.7 An Example Using the Force f, the Length l and the Time t as Basic Dimensions

This is an example using the force f, the length l and time t as basic dimensions, [6]. At the end of a wire of length l_w a mass $m_{a,w}$ is mounted. The mass on the wire rotates around a fixed point with velocity v_w. What is the force f_w in the wire?

As Table 2.7 shows we have $n = 3$ and $r = 3$ and therefore $p = n - r = 0$. This means that no dimensionless power product exists. We recognize immediately from the matrix of dimensions, Table 2.7, that $f_w = const \cdot l_w^{k_1}\, m_{a,w}^{k_2}\, v_w^{k_3}$ must hold and obtain by evaluating the following system of linear equations

$$
\begin{aligned}
0 \cdot k_1 + 1 \cdot k_2 + 0 \cdot k_3 &= 1\,, \\
1 \cdot k_1 - 1 \cdot k_2 + 1 \cdot k_3 &= 0\,, \\
0 \cdot k_1 + 2 \cdot k_2 - 1 \cdot k_3 &= 0\,,
\end{aligned}
\tag{2.56}
$$

the set of coefficients $k_1 = -1$, $k_2 = 1$, $k_3 = 2$ and therefore

$$
f_w = const \cdot \frac{m_{a,w} v_w^2}{l_w}\,, \tag{2.57}
$$

a result which is obvious.

References

1. Buckingham, E.: Contributions to our knowledge of soils. Bulletin 25, USDA Bureau of Soils, Washington (1904)
2. Buckingham, E.: Studies on the movement of soil moisture. Bulletin 38, USDA Bureau of Soils, Washington (1907)
3. Buckingham, E.: On Physically Similar Systems: Illustration of the Use of Dimensional Equations. Phys. Rev. **4**, 345–376 (1914)
4. Buckingham, E.: The principle of similitude. Nature **36**, 396–397 (1915)
5. Zierep, J.: Ähnlichkeitsgesetze und Modellregeln der Strömungsmechanik. G. Braun Verlag, Karlsruhe (1972)
6. Görtler, H.: Dimensionsanalyse, Theorie der physikalischen Dimensionen und Anwendungen. Springer, Berlin (1975)

7. Krause, E.: Strömungslehre, Gasdynamik und aerodynamisches Laboratorium. Teubner Verlag, Stuttgart (2003)
8. Hirschel, E.H.: Basics of Aerothermodynamics, 2nd edn. Springer, Berlin (2015)
9. Schlichting, H., Gersten, K.: Boundary Layer Theory, 8th edn. Springer, Berlin (2000)

Chapter 3
The Fractional Analysis Method

The Fractional Analysis Method is a heuristic method and was developed in the second part of the 19th century by J. W. Strutt, third Baron Rayleigh, 1842–1919, [1], the man who received the Nobel Prize in 1904, see also Sect. 6.16, [2]. It encompasses the relationships of single forces and of single energy parts, which are compared under physical aspects. Of course, in the final outcome there is nothing different to the method of differential equations, Chap. 4, or the dimensional analysis, Chap. 2. Nevertheless it makes sense to have a view on this method in particular from the physical, but also from the historical standpoint. Physically, the reader gets a feeling about the quantities involved. Historically, it shows which interests the scientists had at that time.

John William Strutt, 3rd Baron Rayleigh

3.1 The Rate of Forces

Considering the x-momentum equation in two dimensions, Eq. (4.12) in Chap. 4, we can assign the following proportionalities with respect to forces,[1] [3]

$$\frac{\partial}{\partial t}(\rho u) + \frac{\partial}{\partial x}(\rho u^2 + p) + \frac{\partial}{\partial y}(\rho u v) = \frac{\partial}{\partial x}\tau_{xx} + \frac{\partial}{\partial y}\tau_{xy} + \rho g_x \ . \tag{3.1}$$

[1]prop. means proportional and dim. dimension.

C. Weiland, *Mechanics of Flow Similarities*,
https://doi.org/10.1007/978-3-030-42930-0_3

$$\text{(a)} \quad \frac{\partial}{\partial t}(\rho u) \overset{\text{prop.}}{\Longrightarrow} \frac{\bar{\rho}\bar{u}}{\bar{t}} \sim F_u \overset{\text{dim.}}{\Longrightarrow} \left[\frac{N}{m^3}\right] \qquad \text{unsteady force per volume}$$

$$\text{(b)} \quad \frac{\partial}{\partial x}(\rho u^2) \overset{\text{prop.}}{\Longrightarrow} \frac{\bar{\rho}\bar{u}^2}{l} \sim F_i \overset{\text{dim.}}{\Longrightarrow} \left[\frac{N}{m^3}\right] \qquad \text{force of inertia per volume}$$

$$\text{(c)} \quad \frac{\partial p}{\partial x} \overset{\text{prop.}}{\Longrightarrow} \frac{\bar{p}}{l} \sim F_p \overset{\text{dim.}}{\Longrightarrow} \left[\frac{N}{m^3}\right] \qquad \text{pressure force per volume}$$

$$\text{(d)} \quad \frac{\partial}{\partial x}\tau_{xx} \overset{\text{prop.}}{\Longrightarrow} \bar{\mu}\frac{\bar{u}}{l^2} \sim F_f \overset{\text{dim.}}{\Longrightarrow} \left[\frac{N}{m^3}\right] \qquad \text{friction force per volume}$$

$$\text{(e)} \quad \rho g_x \overset{\text{prop.}}{\Longrightarrow} \bar{\rho}\bar{g} \sim F_g \overset{\text{dim.}}{\Longrightarrow} \left[\frac{N}{m^3}\right] \qquad \text{gravitational force per volume}$$

Further we can also define

$$\text{(f)} \quad \frac{\bar{\sigma}}{l^2} \sim F_c \overset{\text{dim.}}{\Longrightarrow} \left[\frac{N}{m^3}\right] \qquad \text{capillary force per volume}$$

$$\text{(g)} \quad -\bar{g}\Delta\bar{\rho} \sim F_l \overset{\text{dim.}}{\Longrightarrow} \left[\frac{N}{m^3}\right] \qquad \text{lift force per volume}$$

The reference values are $\bar{\rho}$ density, $\bar{u}$ velocity, $\bar{p}$ pressure, $\bar{\mu}$ dynamic viscosity, $\bar{g}$ gravity, $\bar{\sigma}$ surface tension, l characteristic length, $\bar{t}$ characteristic time.

From the above force quantities we build physically reasonable relations which all include the force of inertia F_i.

$$\text{(1)} \quad \frac{F_i}{F_g} = \frac{\bar{\rho}\bar{u}^2}{l} \cdot \frac{1}{\bar{\rho}\bar{g}} = \frac{\bar{u}^2}{l\bar{g}} = Fr \qquad \text{Froude number}$$

$$\text{(2)} \quad \frac{F_i}{F_f} = \frac{\bar{\rho}\bar{u}^2}{l} \cdot \frac{l^2}{\bar{\mu}\bar{u}} = \frac{\bar{\rho}\bar{u}l}{\bar{\mu}} = Re \qquad \text{Reynolds number}$$

$$\text{(3)} \quad \frac{F_p}{F_i} = \frac{\bar{p}}{l} \cdot \frac{l}{\bar{\rho}\bar{u}^2} = \frac{\bar{p}}{\bar{\rho}\bar{u}^2} = Eu \qquad \text{Euler number}$$

$$\text{(4)} \quad \frac{F_u}{F_i} = \frac{\bar{\rho}\bar{u}}{\bar{t}} \cdot \frac{l}{\bar{\rho}\bar{u}^2} = \frac{l}{\bar{u}\bar{t}} = Str \qquad \text{Strouhal number}$$

$$(5)\quad \frac{F_i}{F_c} = \frac{\bar{\rho}\bar{u}^2}{l} \cdot \frac{l^2}{\bar{\sigma}} = \frac{\bar{\rho}\bar{u}^2 l}{\bar{\sigma}} = We \qquad \text{Weber number}$$

$$(6)\quad \frac{F_l}{F_i} = -\frac{\bar{g}\Delta\bar{\rho}}{1} \cdot \frac{l}{\bar{\rho}\bar{u}^2} = -\frac{\bar{g}l}{\bar{u}^2} \cdot \frac{\Delta\bar{\rho}}{\bar{\rho}} \approx \frac{\bar{g}l}{\bar{u}^2} \cdot \frac{T_w - T_\infty}{T_\infty} = \frac{1}{Re^2} \cdot Gr \tag{3.2}$$

In relation (6) the Grashof number Gr is included, Sect. 6.4, which has the form

$$Gr = \frac{\bar{\rho}^2 l^3 \bar{g}(T_w - T_\infty)}{\bar{\mu}^2 T_\infty} . \tag{3.3}$$

3.2 The Rate of Energies

The energy equation in two dimensions and conservative form reads (Eq. (4.16) in Chap. 4)

$$\begin{aligned} &\frac{\partial}{\partial t}(\rho e) + \frac{\partial}{\partial x}((\rho e + p)u) + \frac{\partial}{\partial y}((\rho e + p)v) = \\ &= \frac{\partial}{\partial x}\left(\lambda \frac{\partial T}{\partial x}\right) + \frac{\partial}{\partial y}\left(\lambda \frac{\partial T}{\partial y}\right) + \\ &+ \frac{\partial u \tau_{xx}}{\partial x} + \frac{\partial u \tau_{xy}}{\partial y} + \frac{\partial v \tau_{yx}}{\partial x} + \frac{\partial v \tau_{yy}}{\partial y} + q_H . \end{aligned} \tag{3.4}$$

The lefthand side of Eq. (3.4) can be reformulated, which is helpful for the definition of some of the energy components (heat fluxes), which we will consider below.

$$\begin{aligned} &\frac{\partial}{\partial t}\left(\rho\left(e_v + \frac{1}{2}v^2\right)\right) + \frac{\partial}{\partial x}\left(\left(\rho\left(e_v + \frac{1}{2}v^2\right) + p\right)u\right) \\ &+ \frac{\partial}{\partial y}\left(\left(\rho\left(e_v + \frac{1}{2}v^2\right) + p\right)v\right) = \\ &= \rho\, c_p \frac{\partial T}{\partial t} + \rho\, c_p \mathbf{v} \cdot grad\ T - \frac{\partial p}{\partial t} - \mathbf{v} \cdot grad\ p = \\ &= \rho\, c_p \frac{\partial T}{\partial t} + \rho\, c_p\left(u \frac{\partial T}{\partial x} + v \frac{\partial T}{\partial y}\right) - \frac{dp}{dt} , \end{aligned} \tag{3.5}$$

where the following relations are used

$$h = e_v + \frac{p}{\rho} = c_p T , \tag{3.6}$$

$$\frac{\partial}{\partial t}\left(\frac{1}{2}v^2\right) + \mathbf{v}\cdot grad\,\left(\frac{1}{2}v^2\right) + \frac{1}{\rho}\mathbf{v}\cdot grad\;p = 0\,. \tag{3.7}$$

Now we have instead of Eq. (3.4)

$$\begin{aligned} &\rho\, c_p \frac{\partial T}{\partial t} + \rho\, c_p \left(u\frac{\partial T}{\partial x} + v\frac{\partial T}{\partial y}\right) - \frac{dp}{dt} = \\ = \;& \frac{\partial}{\partial x}\left(\lambda\frac{\partial T}{\partial x}\right) + \frac{\partial}{\partial y}\left(\lambda\frac{\partial T}{\partial y}\right) + \\ + \;& \frac{\partial u\tau_{xx}}{\partial x} + \frac{\partial u\tau_{xy}}{\partial y} + \frac{\partial v\tau_{yx}}{\partial x} + \frac{\partial v\tau_{yy}}{\partial y} + q_H\,. \end{aligned} \tag{3.8}$$

The quantity q_H contains two contributions in our case, namely $q_H = \dot{e}_r + \dot{e}$, where $\dot{e}_r$ denotes the heat flux by radiation and $\dot{e}$ the heat flux by sources (e.g. heat production by chemical reactions). The term dp/dt represents the power of compression due to pressure. Next we proceed in the same way as we did for the forces in the momentum Eq. (3.1), Sect. 3.1. By a look at Eq. (3.8) we can identify the following elements of energy (heat flux elements).

(a) $\rho\, c_p \dfrac{\partial T}{\partial t} \overset{\text{prop.}}{\Longrightarrow} \dfrac{\bar{\rho}\,\bar{c_p}\,\bar{T}}{\bar{t}} \sim \dot{e}_u \overset{\text{dim.}}{\Longrightarrow} \left[\dfrac{N}{m^2 s}\right]$

local heat flux per volume ,

(b) $\rho\, c_p \left(u\dfrac{\partial T}{\partial x} + v\dfrac{\partial T}{\partial y}\right) \overset{\text{prop.}}{\Longrightarrow} \dfrac{\bar{\rho}\,\bar{c_p}\,\bar{T}\,\bar{u}}{l} \sim \dot{e}_c \overset{\text{dim.}}{\Longrightarrow} \left[\dfrac{N}{m^2 s}\right]$

convective heat flux per volume ,

(c) $\dfrac{\partial}{\partial x}\left(\lambda\dfrac{\partial T}{\partial x}\right) + \dfrac{\partial}{\partial y}\left(\lambda\dfrac{\partial T}{\partial y}\right) \overset{\text{prop.}}{\Longrightarrow} \dfrac{\bar{\lambda}\,\bar{T}}{l^2} \sim \dot{e}_\lambda \overset{\text{dim.}}{\Longrightarrow} \left[\dfrac{N}{m^2 s}\right]$

conductive heat flux per volume ,

(d) $\dfrac{\partial u\tau_{xx}}{\partial x} + \dfrac{\partial u\tau_{xy}}{+} \cdots \overset{\text{prop.}}{\Longrightarrow} \dfrac{\bar{\mu}\,\bar{u}^2}{l^2} \sim \dot{e}_f \overset{\text{dim.}}{\Longrightarrow} \left[\dfrac{N}{m^2 s}\right]$

heat flux of friction per volume ,

(e) $\dfrac{\varepsilon\bar{\sigma}\,\bar{T}^4}{l} \sim \dot{e}_r \overset{\text{dim.}}{\Longrightarrow} \left[\dfrac{N}{m^2 s}\right]$

heat flux by radiation per volume ,

(f) $\dot{Q} \sim \dot{e} \overset{\text{dim.}}{\Longrightarrow} \left[\frac{N}{m^2 s}\right]$

heat flux by sources per volume .

Most of the reference quantities are the same as in Sect. 3.1, but in addition we have to list $\bar{c}_p$, the specific heat at constant pressure, $\bar{T}$, the temperature, $\bar{\lambda}$, the thermal conductivity and $\bar{\sigma}$, the Stefan–Boltzmann constant.[2] By physical considerations we can formulate the subsequent proportionalities.

$$(1) \quad \frac{\dot{e}_c}{\dot{e}_\lambda} = \frac{\bar{\rho}\,\bar{c}_p\,\bar{T}\,\bar{u}}{l} \cdot \frac{l^2}{\bar{\lambda}\bar{T}} = \frac{\bar{\rho}\,\bar{c}_p\,\bar{u}\,l}{\bar{\lambda}} = \frac{\bar{u}l}{\bar{k}} = Pe \qquad \text{Peclet number ,}$$

$$(2) \quad \frac{\dot{e}_f}{\dot{e}_c} = \frac{\bar{\mu}\,\bar{u}^2}{l^2} \cdot \frac{l}{\bar{\rho}\,\bar{c}_p\,\bar{T}\,\bar{u}} = \frac{\bar{u}^2}{\bar{c}_p\,\bar{T}} \cdot \frac{1}{Re} = Ec \cdot \frac{1}{Re} \qquad \text{Eckert number ,}$$

$$(3) \quad \frac{\dot{e}_\lambda}{\dot{e}_u} = \frac{\bar{\lambda}\,\bar{T}}{l^2} \cdot \frac{\bar{t}}{\bar{\rho}\,\bar{c}_p\,\bar{T}} = \frac{\bar{k}\,\bar{t}}{l^2} = Fo \qquad \text{Fourier number ,}$$

$$(4) \quad \frac{\dot{e}_r}{\dot{e}_\lambda} = \frac{\varepsilon\,\bar{\sigma}\,\bar{T}^4}{l} \cdot \frac{l^2}{\bar{\lambda}\,\bar{T}} = \frac{\varepsilon\,\bar{\sigma}\,\bar{T}^3\,l}{\bar{\lambda}} = Ste \qquad \text{Stefan number ,}$$

$$(5) \quad \frac{\dot{e}}{\dot{e}_u} = \frac{\bar{\rho}\,\dot{Q}\,\bar{t}}{\bar{\rho}\,\bar{c}_p\,\bar{T}} = \frac{Q}{\bar{c}_p\,\bar{T}} = DAM2 \qquad \text{2. Damköhler number .} \tag{3.9}$$

References

1. Rayleigh, J.W.: On the questions of the stability of the flows of liquids. Phil. Mag. **34**, 59–70 (1892)
2. Lindsay, R.B.: John William Strutt, 3rd Lord Rayleigh. https://www.britannica.com/print/article/492464
3. Zierep, J.: Ähnlichkeitsgesetze und Modellregeln der Strömungsmechanik. G. Braun Verlag, Karlsruhe (1972)

[2] $\bar{\sigma} = 5.688 \cdot 10^{-8} W/m^2 K^4$, ε emission coefficient.

Chapter 4
Method of Differential Equations

First we state that the governing equations, which describe general three dimensional flow fields, can be found in more or less all fluiddynamic textbooks. Some of them give the complete mathematical derivations of the equations both in integral and differential form, [1–6].

There are several different formulations of the equations in differential notation, which encompass the set of dependent variables, namely

- vector notation,
- tensorial notation,
- component presentation:
  - for various forms of dependent variables: conservative and all forms of non-conservative,
  - for different coordinate systems (Cartesian, non-orthogonal curvilinear, others).

For completeness we first list below the governing equations in integral form (conservation of mass, momentum and energy). In a second step we write down the differential equations in component notation with conservative variables, which we strictly use for the following considerations.

C. Weiland, *Mechanics of Flow Similarities*,
https://doi.org/10.1007/978-3-030-42930-0_4

4.1 The Fluid Dynamic Equations in Integral Form

4.1.1 The Conservation of Mass

$$\frac{\partial}{\partial t}\int_V \rho dV + \oint_F \rho \mathbf{v} \cdot d\mathbf{F} = 0 \quad (4.1)$$

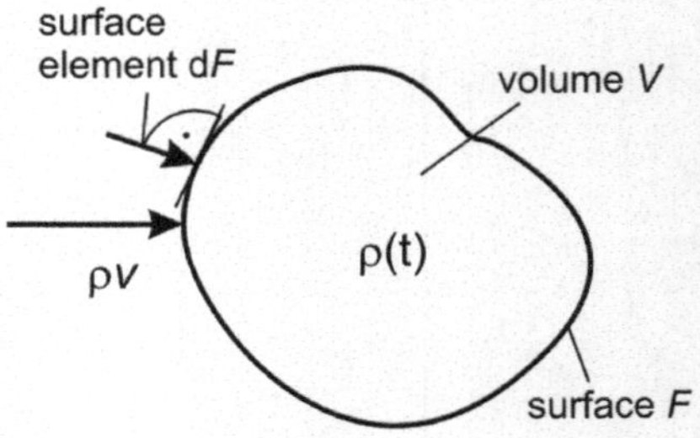

Sketch of a control volume.

4.1.2 The Conservation of Momentum

$$\frac{\partial}{\partial t}\int_V \rho \mathbf{v} dV + \oint_F \rho \mathbf{v}(\mathbf{v} \cdot d\mathbf{F}) = \int_V \rho \mathbf{f}_e dV + \oint_F \bar{\bar{\tau}} d\mathbf{F} - \oint_F p d\mathbf{F} \tag{4.2}$$

4.1.3 The Conservation of Energy

$$\frac{\partial}{\partial t}\int_V \rho e dV + \oint_F \rho e \mathbf{v} \cdot d\mathbf{F} = \oint_F \mathbf{q} \cdot d\mathbf{F} + \int_V (\rho \mathbf{f}_e \cdot \mathbf{v} + q_H) dV + \oint_F ((\bar{\bar{\tau}} - p\bar{\bar{I}}) \cdot \mathbf{v}) \cdot d\mathbf{F} \tag{4.3}$$

The denotation of the variables used in Eqs. (4.1)–(4.3) is as follows: ρ density, $\mathbf{v}$ velocity vector, dV volume element, $d\mathbf{F}$ vector of surface element, t time, p pressure, $\mathbf{f}_e$ vector of external forces per unit mass like gravitational, electric and magnetic ones, etc., $\bar{\bar{\tau}}$ viscous stress tensor, e total energy per unit mass, $\mathbf{q}$ vector of molecular transport of energy, q_H heat sources.

4.2 The Fluid Dynamic Equations in Differential Form

To derive from the integral versions of the governing Eqs. (4.1)–(4.3) the differential ones requires that all the integrands are differentiable, which means that the flow fields to be determined can not contain singularities like shocks, shear layers, etc..

This is different to the integral versions, which are valid also for flows including such singularities.

4.2.1 The Continuity Equation

From Eq. (4.1) we obtain with Gauss' theorem the vector form of the continuity equation

$$\frac{\partial \rho}{\partial t} + div(\rho \mathbf{v}) = 0 \,, \tag{4.4}$$

and the component version then reads

$$\frac{\partial \rho}{\partial t} + \frac{\partial \rho u}{\partial x} + \frac{\partial \rho v}{\partial y} + \frac{\partial \rho w}{\partial z} = 0 \,. \tag{4.5}$$

4.2.2 The Momentum Equation

In the same way as for the continuity equation, Sect. 4.2.1, we find the vector/tensor form of the momentum equation

$$\frac{\partial \rho \mathbf{v}}{\partial t} + div(\rho \mathbf{v} \otimes \mathbf{v} + p\bar{\bar{I}} - \bar{\bar{\tau}}) = \rho \mathbf{f}_e \,, \tag{4.6}$$

where $\mathbf{v} \otimes \mathbf{v}$ denotes a tensor product. For the component version we obtain

$$\begin{aligned}
&\text{x - component} \quad && \frac{\partial}{\partial t}(\rho u) + \frac{\partial}{\partial x}(\rho u^2 + p) + \frac{\partial}{\partial y}(\rho u v) + \frac{\partial}{\partial z}(\rho u w) = \\
& && \frac{\partial}{\partial x}\tau_{xx} + \frac{\partial}{\partial y}\tau_{xy} + \frac{\partial}{\partial z}\tau_{xz} + \rho \mathbf{f}_{e,x} \,, \\
&\text{y - component} \quad && \frac{\partial}{\partial t}(\rho v) + \frac{\partial}{\partial x}(\rho v u) + \frac{\partial}{\partial y}(\rho v^2 + p) + \frac{\partial}{\partial z}(\rho v w) = \\
& && \frac{\partial}{\partial x}\tau_{yx} + \frac{\partial}{\partial y}\tau_{yy} + \frac{\partial}{\partial z}\tau_{yz} + \rho \mathbf{f}_{e,y} \,, \\
&\text{z - component} \quad && \frac{\partial}{\partial t}(\rho w) + \frac{\partial}{\partial x}(\rho w u) + \frac{\partial}{\partial y}(\rho w v) + \frac{\partial}{\partial z}(\rho w^2 + p) = \\
& && \frac{\partial}{\partial x}\tau_{zx} + \frac{\partial}{\partial y}\tau_{zy} + \frac{\partial}{\partial z}\tau_{zz} + \rho \mathbf{f}_{e,z} \,.
\end{aligned} \tag{4.7}$$

The components of the viscous stress tensor are defined by[1]

$$
\begin{aligned}
\tau_{xx} &= \mu \left\{ 2\frac{\partial u}{\partial x} - \frac{2}{3}\left(\frac{\partial u}{\partial x} + \frac{\partial v}{\partial y} + \frac{\partial w}{\partial z} \right) \right\} , \\
\tau_{yy} &= \mu \left\{ 2\frac{\partial v}{\partial y} - \frac{2}{3}\left(\frac{\partial u}{\partial x} + \frac{\partial v}{\partial y} + \frac{\partial w}{\partial z} \right) \right\} , \\
\tau_{zz} &= \mu \left\{ 2\frac{\partial w}{\partial z} - \frac{2}{3}\left(\frac{\partial u}{\partial x} + \frac{\partial v}{\partial y} + \frac{\partial w}{\partial z} \right) \right\} ,
\end{aligned}
\tag{4.8}
$$

$$
\begin{aligned}
\tau_{xy} = \tau_{yx} &= \mu \left(\frac{\partial u}{\partial y} + \frac{\partial v}{\partial x} \right) , \\
\tau_{yz} = \tau_{zy} &= \mu \left(\frac{\partial v}{\partial z} + \frac{\partial w}{\partial y} \right) , \\
\tau_{zx} = \tau_{xz} &= \mu \left(\frac{\partial w}{\partial x} + \frac{\partial u}{\partial z} \right) .
\end{aligned}
\tag{4.9}
$$

Equation (4.7) together with Eqs. (4.8) and (4.9) are the classical equations of motion. Despite the fact that these equations obviously were published independently by the authors Claude Louis Marie Henri Navier (1822), Siméon Denis Poisson (1831), Barré de Saint-Venant (1843) and George Gabriel Stokes (1845), they are named today Navier–Stokes equations.

C.L.M.H. Navier

G.G. Stokes

[1]We have made use of the assumption, that the bulk viscosity can be neglected compared to the dynamic viscosity, [1], which is the same conclusion as to say, that Stokes relation is valid $2\mu + 3\lambda = 0$ with λ being the second viscosity component, [6].

4.2.3 The Energy Equation

The vector form of the energy equation reads

$$\frac{\partial \rho e}{\partial t} + div(\rho\, e\, \mathbf{v}) = div(\mathbf{q}) + div((\bar{\bar{\tau}} - p\bar{\bar{I}}) \cdot \mathbf{v}) + \rho \mathbf{f}_e \cdot \mathbf{v} + q_H \,. \tag{4.10}$$

In components, when we consider for $\mathbf{q}$ only the heat conduction ($\mathbf{q} = \lambda\, grad\, T$), we find

$$\begin{aligned}
&\frac{\partial}{\partial t}(\rho e) + \frac{\partial}{\partial x}((\rho e + p)u) + \frac{\partial}{\partial y}((\rho e + p)v) + \frac{\partial}{\partial z}((\rho e + p)w) = \\
= \ & \frac{\partial}{\partial x}\left(\lambda \frac{\partial T}{\partial x}\right) + \frac{\partial}{\partial y}\left(\lambda \frac{\partial T}{\partial y}\right) + \frac{\partial}{\partial z}\left(\lambda \frac{\partial T}{\partial z}\right) + \\
+ \ & \frac{\partial u \tau_{xx}}{\partial x} + \frac{\partial u \tau_{xy}}{\partial y} + \frac{\partial u \tau_{xz}}{\partial z} + \frac{\partial v \tau_{yx}}{\partial x} + \frac{\partial v \tau_{yy}}{\partial y} + \frac{\partial v \tau_{yz}}{\partial z} + \\
+ \ & \frac{\partial w \tau_{zx}}{\partial x} + \frac{\partial w \tau_{zy}}{\partial y} + \frac{\partial w \tau_{zz}}{\partial z} + \rho(f_{e,x}u + f_{e,y}v + f_{e,z}w) + q_H \,.
\end{aligned} \tag{4.11}$$

with T being the temperature, λ the thermal conductivity, $\rho \mathbf{f}_e \cdot \mathbf{v}$ the power of the external forces, $e = e_v + 1/2v^2$ the total energy per unit mass, e_v the internal energy per unit mass, $1/2v^2$ the kinetic energy per unit mass. For our considerations we neglect $\mathbf{f}_e$ in the energy equation.

Since scientists began to solve numerically the governing equations for three dimensional flow fields, namely the equations for continuity Eq. (4.5), of motion Eq. (4.7) and of energy Eq. (4.11), this set of five partial differential equations in the related literature was summarily called Navier–Stokes equations.

4.3 Dimensionless Forms of the Governing Equations

4.3.1 The Dimensionless Momentum Equation

To demonstrate the procedure for making the momentum equation, Eq. (4.7), dimensionless, we consider its x-component in two dimensions, where we include the gravitation vector $\mathbf{g}$ as component of the external force $\mathbf{f}$, [7].

$$\frac{\partial}{\partial t}(\rho u) + \frac{\partial}{\partial x}(\rho u^2 + p) + \frac{\partial}{\partial y}(\rho u v) = \frac{\partial}{\partial x}\tau_{xx} + \frac{\partial}{\partial y}\tau_{xy} + \rho g_x \tag{4.12}$$

It is a special skill to define the reference values. The choice of these quantities are mostly adapted to the configurational flow problem and/or to the degree of the approximation of the governing equations considered. In the general case we start with the definitions

$$u' = \frac{u}{\bar{v}},\ v' = \frac{v}{\bar{v}},\ \rho' = \frac{\rho}{\bar{\rho}},\ p' = \frac{p}{\Delta\bar{p}},\ g'_x = \frac{\rho g_x}{\bar{\rho} g},$$
$$t' = \frac{t}{\bar{t}},\ x' = \frac{x}{l},\ y' = \frac{y}{l},\ \mu' = \frac{\mu}{\bar{\mu}}. \tag{4.13}$$

Introducing all that in Eq. (4.12) we obtain the dimensionless form of the x-momentum equation

$$\frac{l}{\bar{t}\bar{v}}\frac{\partial}{\partial t'}(\rho' u') + \frac{\partial}{\partial x'}(\rho' u'^2) + \frac{\partial}{\partial y'}(\rho' u' v') + \frac{\Delta\bar{p}}{\bar{\rho}\bar{v}^2}\frac{\partial p'}{\partial x'} =$$
$$= \frac{\bar{\mu}}{l\bar{\rho}\bar{v}}\frac{\partial}{\partial x'}\tau'_{xx} + \frac{\bar{\mu}}{l\bar{\rho}\bar{v}}\frac{\partial}{\partial y'}\tau'_{yx} + \frac{gl}{\bar{v}^2}g'_x\,, \tag{4.14}$$

and we then can identify the Strouhal number Str, the Euler number Eu, the Reynolds number Re and the Froude number Fr

$$Str\frac{\partial}{\partial t'}(\rho' u') + \frac{\partial}{\partial x'}(\rho' u'^2) + \frac{\partial}{\partial y'}(\rho' u' v') + Eu\frac{\partial p'}{\partial x'} =$$
$$= \frac{1}{Re}\frac{\partial}{\partial x'}\tau'_{xx} + \frac{1}{Re}\frac{\partial}{\partial y'}\tau'_{yx} + \frac{1}{Fr}g'_x\,. \tag{4.15}$$

4.3.2 The Dimensionless Energy Equation

Also for the energy equation, Eq. (4.11), for our purposes it is sufficient to consider its two dimensional version

$$\frac{\partial}{\partial t}(\rho e) + \frac{\partial}{\partial x}((\rho e + p)u) + \frac{\partial}{\partial y}((\rho e + p)v) =$$
$$= \frac{\partial}{\partial x}\left(\lambda\frac{\partial T}{\partial x}\right) + \frac{\partial}{\partial y}\left(\lambda\frac{\partial T}{\partial y}\right) +$$
$$+ \frac{\partial u\tau_{xx}}{\partial x} + \frac{\partial u\tau_{xy}}{\partial y} + \frac{\partial v\tau_{yx}}{\partial x} + \frac{\partial v\tau_{yy}}{\partial y} + q_H\,. \tag{4.16}$$

To make Eq. (4.16) dimensionless we apply the following relations

$$u' = \frac{u}{\bar{v}},\ v' = \frac{v}{\bar{v}},\ \rho' = \frac{\rho}{\bar{\rho}},\ p' = \frac{p}{\bar{\rho}\bar{v}^2},\ \lambda' = \frac{\lambda}{\bar{\lambda}},$$
$$t' = \frac{t}{\bar{t}},\ x' = \frac{x}{l},\ y' = \frac{y}{l},\ \mu' = \frac{\mu}{\bar{\mu}},\ T' = \frac{T}{\bar{T}},\ e' = \frac{e}{\bar{v}^2}\ , \tag{4.17}$$

and obtain

$$\begin{aligned}&\frac{l}{\bar{v}\bar{t}}\frac{\partial}{\partial t'}(\rho' e') + \frac{\partial}{\partial x'}(\rho' e' + p')u' + \frac{\partial}{\partial y'}(\rho' e' + p')v' = \\ &= \frac{\bar{\mu}}{l\bar{v}\bar{\rho}}\left(\frac{\partial}{\partial x'}(u'\tau'_{xx} + v'\tau'_{xy}) + \frac{\partial}{\partial y'}(u'\tau'_{yx} + v'\tau'_{yy})\right) + \\ &+ \frac{\bar{\lambda}\bar{T}}{l\bar{\rho}\bar{v}^3}\left(\frac{\partial}{\partial x'}\left(\lambda'\frac{\partial T'}{\partial x'}\right) + \frac{\partial}{\partial y'}\left(\lambda'\frac{\partial T'}{\partial y'}\right)\right) + q_H\ .\end{aligned} \tag{4.18}$$

The coefficient at the time derivative again is the Strouhal number $Str = l/\bar{v}\bar{t}$ and the one at the viscous stress tensor components the Reynolds number $Re = l\bar{v}\bar{\rho}/\bar{\mu}$. The heat conduction terms have the coefficient $\bar{\lambda}\bar{T}/l\bar{\rho}\bar{v}^3$. With the thermal diffusivity $\bar{k} = \bar{\lambda}/\bar{c_p}\bar{\rho}$ we obtain

$$\frac{\bar{\lambda}\bar{T}}{l\bar{\rho}\bar{v}^3} = \frac{\bar{k}\bar{\rho}\bar{c_p}\bar{T}}{l\bar{\rho}\bar{v}^3} = \frac{\bar{k}}{l\bar{v}}\cdot\frac{\bar{c_p}\bar{T}}{\bar{v}^2} = \frac{1}{Pe}\cdot\frac{1}{(\gamma-1)M^2}\ , \tag{4.19}$$

where Pe denotes the Pèclet number, $M^2 = \bar{v}^2/(\bar{c}_p\bar{T}(\gamma-1)) = \bar{v}^2/\bar{a}^2$ the Mach number and γ the ratio of specific heats. The Reynolds number Re and the Pèclet number Pe are connected by the Prandtl number Pr.

$$\frac{l\bar{\rho}\bar{v}\bar{c_p}}{\bar{\lambda}}\cdot\frac{\bar{\mu}}{l\bar{\rho}\bar{v}} = \frac{l\bar{v}}{\bar{k}}\cdot\frac{\bar{\nu}}{l\bar{v}} = \frac{\bar{\nu}}{\bar{k}} = \frac{Pe}{Re} = Pr\ , \tag{4.20}$$

with $\bar{\nu} = \bar{\mu}/\bar{\rho}$ the kinematic viscosity.

In case that we consider a high temperature gas flow, where non-equilibrium real-gas effects are present, the molecular transport of energy **q** has to be extended by the thermal transport due to mass diffusion, [1]. We then define for the x-component of **q** of a binary gas mixture

$$\frac{\partial}{\partial x}q_x = \frac{\partial}{\partial x}\left(\lambda\frac{\partial T}{\partial x} + \rho D_{AB}\sum_{\alpha=A,B} c_{p,\alpha}\, T\,\frac{\partial \omega_\alpha}{\partial x}\right)\ . \tag{4.21}$$

where D_{AB} denotes the binary mass diffusivity coefficient, $c_{p,\alpha}$ the specific heat at constant pressure and ω_α the mass fraction in both cases of the αth species, [1].

Non-dimensioning Eq. (4.21) and introducing this relation in the environment of Eq. (4.18) leads to

$$\frac{\bar{\lambda}\bar{T}}{l\bar{\rho}\bar{v}^3}\frac{\partial}{\partial x'}\left(\lambda'\frac{\partial T'}{\partial x'}\right)+\frac{\bar{D}_{AB}\,\bar{c}_p\,\bar{T}}{l\bar{v}^3}\frac{\partial}{\partial x'}\left(\rho' D'_{AB}\sum_{\alpha=A,B} c'_{p,\alpha}T'\frac{\partial\omega_\alpha}{\partial x'}\right), \tag{4.22}$$

where we add $D'_{AB} = D_{AB}/\bar{D}_{AB}$ and $c'_{p,\alpha} = c_{p,\alpha}/\bar{c}_p$ to the relations (4.17).

The coefficient left of the first differential operator in relation Eq. (4.22) was already evaluated (Eq. (4.19)), the coefficient left of the second differential operator yields

$$\frac{\bar{D}_{AB}\,\bar{c}_p\,\bar{T}}{l\bar{v}^3}=\frac{\bar{\lambda}\bar{T}}{l\bar{\rho}\bar{v}^3}\cdot\frac{\bar{\rho}\,\bar{D}_{AB}\,\bar{c}_p}{\bar{\lambda}}=\frac{1}{Pe\,(\gamma-1)M^2}\cdot\frac{1}{Le}\,. \tag{4.23}$$

Here

$$Le=\frac{\bar{\lambda}}{\bar{\rho}\,\bar{c}_p\,\bar{D}_{AB}}=\frac{\bar{k}}{\bar{D}_{AB}} \tag{4.24}$$

is the Lewis number.

4.4 Derivation of the Boundary Layer Equations

In the following the boundary layer equations are derived from the Navier–Stokes equations using the dimensional analysis calculus. The general time-dependent Navier–Stokes equations in three dimensions are given by the Eqs. (4.5), (4.7) and (4.11). In order to make the derivation of the boundary layer equations as clearly as possible, the two dimensional incompressible Navier–Stokes equations (continuity equation and momentum equations) are utilized, [8].

$$\begin{aligned}\frac{\partial u}{\partial x}+\frac{\partial v}{\partial y}&=0\,,\\ u\frac{\partial u}{\partial x}+v\frac{\partial u}{\partial y}&=-\frac{1}{\rho}\frac{\partial p}{\partial x}+\nu\Delta u\,,\\ u\frac{\partial v}{\partial x}+v\frac{\partial v}{\partial y}&=-\frac{1}{\rho}\frac{\partial p}{\partial y}+\nu\Delta v\,.\end{aligned} \tag{4.25}$$

The dimensionless variables[2]

[2]The density ρ is a constant and also the kinematic viscosity ν, which does not depend on the temperature T.

$$u' = \frac{u}{\bar{v}}, \quad v' = \frac{v}{\bar{v}}, \quad p' = \frac{p}{\rho \bar{v}^2}, \quad x' = \frac{x}{l}, \quad y' = \frac{y}{\delta}$$

are applied and it is assumed that the orders of magnitudes

$$u' \sim 1, \quad p' \sim 1, \quad x' \sim 1, \quad y' \sim 1$$

are valid. From the continuity equation one obtains

$$\frac{\partial u'}{\partial x'} + \frac{\partial v'}{\partial y'} \frac{l}{\delta} = 0 \,, \tag{4.26}$$

which results in the order of magnitude estimation

$$v'' = v' \frac{l}{\delta} \sim 1 \,,$$

if the operators $\partial/\partial x'$ and $\partial/\partial y'$ are not changing the order of magnitude.

Inserting this result in the dimensionless x-momentum equation yields

$$\begin{aligned} u' \frac{\partial u'}{\partial x'} + v'' \frac{\partial u'}{\partial y'} &= -\frac{\partial p'}{\partial x'} + \underbrace{\frac{1}{Re} \frac{l^2}{\delta^2}}_{\sim 1} \left\{ \underbrace{\frac{\delta^2}{l^2} \frac{\partial^2 u'}{\partial x'^2}}_{\ll 1} + \underbrace{\frac{\partial^2 u'}{\partial y'^2}}_{\sim 1} \right\} \\ &= -\frac{\partial p'}{\partial x'} + \frac{1}{Re} \frac{l^2}{\delta^2} \left\{ \frac{\partial^2 u'}{\partial y'^2} \right\} \end{aligned} \tag{4.27}$$

and it follows, assumed that at least one of the viscous terms is active and that the convective terms and the pressure term are proportional to unity (~ 1) for $Re \gg 1$

$$\frac{\delta}{l} \sim \frac{1}{\sqrt{Re}} \,, \tag{4.28}$$

where by this condition the length δ can be identified as the boundary layer thickness.

With the relation Eq. (4.28) the dimensionless quantities y' and v'' are rearranged and one has

$$y' = \frac{y}{l} \sqrt{Re} \,, \quad v'' = \frac{v}{\bar{v}} \sqrt{Re} \,. \tag{4.29}$$

The dimensionless y-momentum equation then has the form

$$\frac{1}{Re} \left\{ u' \frac{\partial v''}{\partial x'} + v'' \frac{\partial v''}{\partial y'} \right\} = -\frac{\partial p'}{\partial y'} + \frac{1}{Re} \frac{\partial^2 v''}{\partial y'^2} + \frac{1}{Re^2} \frac{\partial^2 v''}{\partial x'^2} \,. \tag{4.30}$$

For $Re \gg 1$ Eq. (4.30) results in

$$\frac{\partial p'}{\partial y'} = 0\,. \tag{4.31}$$

Equations (4.27) and (4.31) together with (4.26) are the boundary layer equations for steady, two dimensional, incompressible flows.

Remark The momentum equations for steady, three dimensional, compressible flows in the boundary layer approach read, [1]:

$$\begin{aligned}
\rho' u' \frac{\partial u'}{\partial x'} + \rho' v' \frac{\partial u'}{\partial y'} + \rho' w' \frac{\partial u'}{\partial z'} &= -\frac{\partial p'}{\partial x'} + \frac{\partial}{\partial y'}\left(\mu' \frac{\partial u'}{\partial y'}\right), \\
\frac{\partial p'}{\partial y'} &= 0\,, \\
\rho' u' \frac{\partial w'}{\partial x'} + \rho' v' \frac{\partial w'}{\partial y'} + \rho' w' \frac{\partial w'}{\partial z'} &= -\frac{\partial p'}{\partial z'} + \frac{\partial}{\partial y'}\left(\mu' \frac{\partial w'}{\partial y'}\right).
\end{aligned} \tag{4.32}$$

References

1. Hirschel, E.H.: Basics of Aerothermodynamics, 2nd edn. Springer, Berlin (2015)
2. Hirschel, E.H., Weiland, C.: Selected Aerothermodynamic Design Problems of Hypersonic Flight Vehicles. Springer, Berlin (2009) (Progress in Astronautics and Aeronautics, vol. 229, AIAA, Reston, Va.)
3. Becker, E.: Gasdynamik. Teubner Verlag, Stuttgart (1966)
4. Zierep, J.: Theoretische Gasdynamik. G. Braun Verlag, Karlsruhe (1972)
5. Bird, R.B., Stuart, W.E., Lightfoot, E.N.: Transport Phenomena, 2nd edn. Wiley, New York (2002)
6. Hirsch, C.: Numerical Computation of Internal and External Flows, vol. I. Wiley, New York, Reprint (1997)
7. Krause, E.: Strömungslehre, Gasdynamik und aerodynamisches Laboratorium. Teubner Verlag, Stuttgart (2003)
8. Zierep, J.: Ähnlichkeitsgesetze und Modellregeln der Strömungsmechanik. G. Braun Verlag, Karlsruhe (1972)

Chapter 5
Classification of Dimensionless Numbers—Similarity Parameters

This chapter gives an overview over the dimensionless numbers[1] treated in this book, whereat the details of these quantities are investigated in Chap. 6. The basic equations for the aerodynamic, aerothermodynamic and all sorts of heat transfer problems are the Navier–Stokes equations with the appropriate initial and boundary conditions and if necessary expanded by the relations for the thermodynamic equilibrium state and in particular by the equations for the thermodynamic non-equilibrium state (species continuity, excitation of molecular vibrations, electron modes) [1].

Most of the dimensionless numbers examined in this book did originate from considerations of these equations, see Chaps. 3 and 4. Of course all these numbers can also be derived by the pure mathematical examination of the Π theorem of Buckingham, Chap. 2 [2]. Therefore we have arranged some of these dimensionless numbers with respect to the above mentioned equations. Further there are other commonalities for the remaining dimensionless numbers, which are listed, too.

The dimensionless numbers, see Sects. 5.1–5.6, have their major relevance and also their basic application for specific disciplines. In the following we have a view at this aspect.[2]

- *Aerodynamics, aerothermodynamics, gasdynamics, fluid mechanics*:
  Reynolds number, Mach number, Stanton number, Prandtl number, Strouhal number, ratio of specific heats, Euler number in form of the pressure coefficient c_p, (wall temperature parameter), (binary scaling parameter), (Stefan number), (Knudsen number), (Eckert number), (1. Damköhler number).
- *All sorts of heat transfer including free and forced convection*:
  Fourier number, Grashof number, Nusselt number, Rayleigh number, Reynolds number, (Lewis number), (Schmidt number), (Eckert number), (Pèclet number).

---

[1] Many of the quantities listed here can be named both as dimensionless numbers and as similarity parameters. To shorten it we call in the following these only dimensionless numbers.

[2] The dimensionsless numbers in brackets are not often used.

C. Weiland, *Mechanics of Flow Similarities*,
https://doi.org/10.1007/978-3-030-42930-0_5

- *Chemical and process engineering (fluid and gas mixture processes)*:
  Nusselt number, Grashof number, Rayleigh number, Lewis number, Schmidt number, Prandtl number, 2. Damköhler number.
- *Water waves and ship design*:
  Froude number, Reynolds number.
- *Creeping flows*:
  Stokes number, Reynolds number, Euler number.
- *Evolution of spray structures and capillary flows*:
  Weber number.

5.1 Forces—Momentum Equations

Table 5.1 Dimensionless numbers based on the momentum equations

	Names	Lifetime	Equations	Explanations
	L. Euler	1707–1783	$Eu = \frac{p}{\rho u^2}$	Ratio of pressure force to inertial force
	W. Froude	1810–1879	$Fr = \frac{u^2}{gl}$	Ratio of inertial force to gravitational force
	M. Weber	1871–1951	$We = \frac{\rho u^2 l}{\sigma}$	Ratio of inertial force to capillary force
	F. Grashof	1826–1893	$Gr = \frac{g\beta l^3 \Delta T}{\nu^2}$	Ratio of lift force to inertial force $\cdot$ Re^2
	O. Reynolds	1842–1912	$Re = \frac{ul}{\nu}$	Ratio of inertial force to friction force

5.2 Energy Balance—Energy Equation

Table 5.2 Dimensionless numbers based on the energy equation

	Names	Lifetime	Equations	Explanations
	J.C.F. Pèclet	1793–1857	$Pe = \frac{ul}{k} = Re \cdot Pr$	Ratio of heat convection to heat conduction
	E.R.G. Eckert	1904–2004	$Ec = \frac{u^2}{c_p \Delta T}$	Ratio of kinetic energy to enthalpy difference
	G. Damköhler	1908–1944	$DAM2 = \frac{Q}{c_p \Delta T}$	Ratio of energy of thermo-chemical reactions to enthalpy difference
	E.K.W. Nusselt	1882–1957	$Nu = \frac{\alpha l}{\lambda}$	Ratio of heat transfer to heat conduction
	T.E. Stanton	1865–1931	$St = \frac{\alpha}{\rho u c_p}$	Ratio of heat transfer to heat convection
	J. Stefan	1835–1893	$Ste = \frac{\varepsilon \sigma T^3 l}{\lambda}$	Ratio of heat radiation to heat conduction

5.3 Time Dependency

Table 5.3 Dimensionless numbers describing unsteady flows

	Names	Lifetime	Equations	Explanations
	J.B.J. Fourier	1768–1830	$Fo = \frac{kt}{l^2}$	Ratio of rate of heat conduction to rate of heat storage
	V. Strouhal	1850–1922	$Str = \frac{l}{ut}$	Characterization of unsteady flows with typical frequencies

5.4 Material Based Dimensionless Numbers

Table 5.4 Material based dimensionless numbers

	Names	Lifetime	Equations	Explanations
	W. Lewis	1882–1975	$Le = \frac{k}{D} = \frac{Sc}{Pr}$	Heat transport by conduction to heat transport by mass diffusion
	E. Schmidt	1892–1975	$Sc = \frac{\nu}{D} = Le \cdot Pr$	Heat transport by friction to heat transport by mass diffusion
	L. Prandtl	1875–1953	$Pr = \frac{\nu}{k}$	Diffusive momentum transport to convective heat transport

5.5 Miscellaneous Dimensional Numbers

Table 5.5 Other dimensionless numbers and/or similarity parameters

	Names	Lifetime	Equations	Explanations
	M. Knudsen	1871–1949	$Kn = \frac{\lambda}{L}$	Ratio of free mean path of molecules to characteristic length
	E. Mach	1838–1916	$M = \frac{u}{a} = \frac{1}{\sqrt{\gamma Eu}}$	Ratio of flow velocity to speed of sound
	G. Damköhler	1908–1944	$DAM1 = \frac{t_{res}}{\tau}$	Ratio of fluid-dynamic residence time to chemical reaction time
	Wall temperature parameter		$\frac{T_w - T_\infty}{T_\infty} = \frac{Gr \cdot Fr}{Re^2}$	Flows with wall temperatures
	Binary scaling parameter		$\rho \cdot L$	Thermodynamic high temperature reactions[a]
	Ratio of specific heats		$\gamma = \frac{c_p}{c_v}$	

[a] $\rho \cdot L$ is not a dimensionless number, but high-enthalpy flow experience has shown, that flows with constant values of this quantity exhibit a similar behavior (binary scaling parameter) [3, 4]

5.6 Composite Dimensionless Numbers

Table 5.6 Compound dimensionless numbers

	Names	Lifetime	Equations	Description
	J.W. Rayleigh	1842–1919	$Ra = Gr \cdot Pr = \frac{g\beta l^3 \Delta T}{\nu k}$	Ratio between friction, buoyancy and heat conduction
	G.G. Stokes	1819–1903	$Sto = Eu \cdot Re = \frac{pl}{\mu u}$	Ratio of pressure force to friction force

5.7 Timelines of the Scientists, Whose Names the Dimensionless Numbers Bear

In Chap. 1 we have already mentioned that there was a great need for the disciplines "analysis of dimensions" and/or "theory of similarity" mainly in the fields of fluid dynamics and thermodynamics. In the 18th century L. Euler has established the governing equations (momentum equations) of the—at that time called—hydrodynamics in differential form for inviscid flows, today known after him as the Euler equations, followed by the heat transport equation (energy equation) introduced by J.B.J. Fourier. Then in the first half of the 19th century the momentum equations for flows with friction were derived by C.L.M.H. Navier, S.B. Poisson, B. de Saint-Venant and G.G. Stokes. These equations form together with the equations for continuity and energy the set of equations today called the Navier–Stokes equations. In Chap. 4 we have listed the Navier–Stokes equations both in integral and differential form.

From that time on more than one hundred years (up to approximately the 60s of the last century) absolutely no chance was there to solve these equations in full. A lot of activities were undertaken to reduce these equations to such an extent that at least for simple or very simple problems an integration could be achieved. The solutions of most of these reduced equations were more or less only of academic relevance. Examples of such reduced equations are the Stokes equations, the Burger's equation [5], the boundary layer equations for similar flow profiles [6], the linear potential equation, etc., where the latter could be solved by the method of singularity (subsonic flow) and graphically by the method of characteristics (supersonic flow) for simple two dimensional problems [7–10]. The relevance of these solutions for engineering applications was rather low.

On the other hand, when people had observed the various forms of flows[3] (e.g.: in rivers and channels, through pipes, in the atmosphere, around any rigid body, etc.), there was a growing interest to understand the phenomena (beyond the direct solution of the Navier–Stokes equations), which are the stimulus of such flows mainly with the goal to control and to govern these.

The physical problems considered in this book are the motion and the thermodynamics of fluids which are in contact with vehicle surfaces, all sorts of pipes and channels, all sorts of propulsion systems and others. In general these problems depend on flow field variables,[4] thermodynamic variables,[5] material coefficients[6] and geometrical relations.[7]

This means that often the considered problem depends on a large number of variables. As we have shown in the chapters before, this number of variables can be reduced with the help of the analysis of dimensions and sometimes, if $p = 0$, even a direct solution up to a constant can be achieved, see Subsect. 2.2.7.

With the results of the analysis of dimensions one obtains a survey about the dependencies of the physical problem considered and often is able to estimate what changes of variables cause. Further with the knowledge of the dimensionless numbers (similarity parameters), which are valid for a specific physical problem, it may be possible to design and fabricate models, which are geometrically similar to the original objects (flight vehicles, reactors for chemical engineering, etc.). These then can be experimentally investigated and the results can be transferred to the original objects, see Chap. 8.

A look at Fig. 5.1 exhibits that most of the scientists and engineers whose names the various dimensionless numbers bear, were active in the time period 1850–1950. Of course L. Euler has lived a century before, but it can be expected that he had not introduced the number, which is named after him. From J.B.J. Fourier we know that he was engaged with the role which dimensions play in mathematical descriptions of physical processes. For example, he had concluded[8] that summands in a physical relation must have the same dimensions. This behavior is today known as the homogeneity of dimensions.

In [2] it is reported that Hermann v. Helmholtz[9] in 1873 had performed investigations with power products of the hydrodynamic equations, where obviously already

[3]The same is true for all sorts of heat transfer effects.

[4]Like the velocity components u, v, w, the pressure p, etc.

[5]Like the density ρ, the temperature T, the speed of sound a, etc.

[6]Like the dynamic viscosity μ, the thermal conductivity λ, the thermal diffusivity k, etc.

[7]Like the length l, the diameter d, surface coordinate triples x_i, y_j, z_k, etc.

[8]In: Analytical Theory of Heat, 1822, [11].

[9]Hermann Ludwig Ferdinand v. Helmholtz was a German physiologist and physicist born on August 31, 1821 in Potsdam, Germany and died on September 8, 1894 in Berlin.

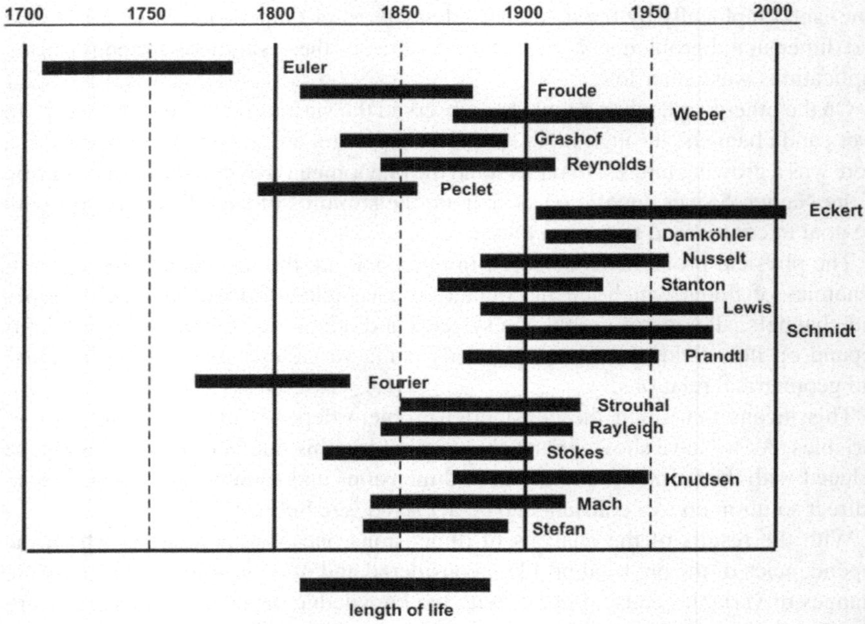

Fig. 5.1 Chronological table and lifetime of the scientists, whose names the dimensionless numbers bear

a variable constellation was identified, which is today known as the Reynolds number (This means 10 years before O. Reynolds had published his oberservations, see Sect. 6.5.).

A systematic formulation of the analysis of dimensions was given with the development of the Π theorem. Several authors, [2, 12, 13], referred to that A. Federmann was the first who had outlined the basic ideas of the Π theorem in 1911 [14], three years before E. Buckingham had published his work [15].

From the mathematical point of view Federmann's work was not completely satisfactory, regarding the quality of the assumptions and the completeness of the mathematical proof. This statement stems from H. Görtler and G.I.Barenblatt. Therefore they presented a detailed and general mathematical proof of the Π theorem in [2, 16].

5.8 The Role of the Dimensionless Numbers—Similarity Parameters

In previous days, when the modern experimental and theoretical (numerical) methods were not yet developed, the roles of the dimensionless numbers and/or similarity parameters were as follows:

1. characterisation of physical regimes,
2. reduction of variables which describe a physical problem,
3. use as variables in mathematical relations and estimations,
4. act as similarity parameters, which enable model tests with the goal to transfer results to the original vehicles, reactors or other technical devices.

Examples regarding 1 are:

- Reynolds number Re, describing laminar, turbulent and transitional flow regimes, Sect. 6.5,
- Mach number M, describing subsonic, transonic, supersonic and hypersonic flow regimes, Sect. 6.19,
- Knudsen number Kn, describing continuum flow, transitional flow, disturbed molecular flow, free molecular flow regimes, Sect. 6.18,
- Rayleigh number Ra: critical Rayleigh number Ra_c characterizes various formations of Rayleigh—Bénard flow systems, Sect. 6.16,
- wall temperature parameter. The wall temperature strongly influences all boundary layer properties including laminar-turbulent transition and turbulence: thermal surface effects, [3].

Comments on 2 are:

For fluid dynamical problems we typically encounter as dimensionless variables the Reynolds number Re, the Mach number M, the Prandtl number Pr, the Froude number Fr, the Strouhal number Str and others.

For heat transfer problems we find as dimensionless variables the Nusselt number Nu, the Rayleigh number Ra, the Grashof number Gr, the Fourier number Fo, the Prandtl number Pr and others, see Chap. 2.

Comments on 3 are:

Some of the dimensionless numbers occur in mathematical estimations, semi-empirical formulas, in the dimensionless forms of the governing equations and other mathematical relations. In the following we give some examples:

- The boundary layer thickness is given by the proportionality $\frac{\delta}{l} \sim \frac{1}{\sqrt{Re_l}}$, the thermal boundary layer thickness by $\frac{\delta_T}{l} \sim \frac{1}{\sqrt{Re_l Pr}}$,
- for the Knudsen number the relation is found $Kn = 1.28\sqrt{\gamma}\frac{M_\infty}{Re_{\infty,l}}$, Sect. 6.18,

- the water wave length generated by a ship relates to $\lambda = 2\pi Frl$, Sect. 6.2,
- the temperature profile due to dissipation in a Couette flow reads $T - T^* = \frac{1}{2} Pr \cdot Ec\, y(1-y)\, T^*$, Sect. 6.7,
- the large amount of semi-empirical formulas of the Nusselt number Nu is essentially given as function of the dimensionless numbers Pr, Re, Gr, (Ra), Sect. 6.9,
- the temperature evolution in a heat conduction process, Sect. 6.14, depends on the Fourier number Fo: $\frac{T(x,t) - T_\infty}{T_i - T_\infty} = \sum_{n=1}^{\infty} C_2 sin\left(\frac{n\pi}{2s}x\right) e^{-(\frac{n\pi}{2})^2 Fo}$,
- the linear Stokes equations with various reference lengths l_x, l_y in x- and y-direction, Eq. (6.42), Sect. 6.17, reads:

$$\left(\frac{\partial^2 u'}{\partial x'^2} + \frac{\partial^2 u'}{\partial y'^2}\left(\frac{l_x}{l_y}\right)^2\right) = -Sto\frac{\partial p'}{\partial x'},$$

$$\left(\frac{l_y}{l_x}\right)^2\left(\frac{\partial^2 v'}{\partial x'^2} + \frac{\partial^2 v'}{\partial y'^2}\left(\frac{l_x}{l_y}\right)^2\right) = -Sto\frac{\partial p'}{\partial y'}.$$

Below we present some sets of dimensionless variables applied to the governing equations and describe their consequences:

The Euler equations and the Navier–Stokes equations, when solved with numerical methods, are applied always in a dimensionless form. There exist some alternatives due to the choice of the set of dimensionless variables. In Subsect. 4.3.1 we used to make the viscous momentum equation dimensionless, Eq. (4.13)

$$\text{set I} \qquad u' = \frac{u}{\bar{v}},\ v' = \frac{v}{\bar{v}},\ \rho' = \frac{\rho}{\bar{\rho}},\ p' = \frac{p}{\Delta\bar{p}},$$
$$t' = \frac{t}{\bar{t}},\ x' = \frac{x}{\bar{l}},\ y' = \frac{y}{\bar{l}},\ \mu' = \frac{\mu}{\bar{\mu}}. \tag{5.1}$$

and found in the dimensionless equation, (Eq. (4.15)), the Strouhal number Str, the Euler number Eu and the Reynolds number Re (Note: The gravitational force **g** per unit mass is neglected in this consideration).

For the energy equation, Subsect. 4.3.2, we applied, Eq. (4.17)

$$\text{set II} \qquad u' = \frac{u}{\bar{v}},\ v' = \frac{v}{\bar{v}},\ \rho' = \frac{\rho}{\bar{\rho}},\ p' = \frac{p}{\bar{\rho}\bar{v}^2},\ \lambda' = \frac{\lambda}{\bar{\lambda}},$$
$$t' = \frac{t}{\bar{t}},\ x' = \frac{x}{\bar{l}},\ y' = \frac{y}{\bar{l}},\ \mu' = \frac{\mu}{\bar{\mu}},\ T' = \frac{T}{\bar{T}},\ e' = \frac{e}{\bar{v}^2}. \tag{5.2}$$

and again found the Strouhal number Str and the Reynolds number Re, as well as further the Prandtl number Pr (or the Peclet number $Pe = Re \cdot Pr$) and the Mach number M, (Eq. (4.18)), and when the equation is extended to chemical non-equilibrium thermodynamics with $D'_{AB} = D_{AB}/\overline{D_{AB}}$ (binary system), $c'_{p,\alpha} = c_{p,\alpha}/\bar{c_p}$ the Lewis number Le (or the Schmidt number $Sc = Le \cdot Pr$, Eq. (4.23)).

Typically for the Navier–Stokes equations (including chemical non-equilibrium thermodynamics) the

$$\text{set III} \qquad \begin{aligned} & u' = \frac{u}{\bar{v}},\ v' = \frac{v}{\bar{v}},\ \rho' = \frac{\rho}{\bar{\rho}},\ p' = \frac{p}{\bar{\rho}\bar{v}^2},\ \lambda' = \frac{\lambda}{\bar{\lambda}}, \\ & t' = \frac{t\bar{v}}{l},\ x' = \frac{x}{l},\ y' = \frac{y}{l},\ \mu' = \frac{\mu}{\bar{\mu}},\ T' = \frac{T}{\bar{T}}, \\ & e' = \frac{e}{\bar{v}^2},\ D'_{AB} = \frac{D_{AB}}{\overline{D_{AB}}}. \end{aligned} \tag{5.3}$$

is used, which differs from set II by the definition of the dimensionless time t', whereby the Strouhal number is dropped in the dimensionless equations, see for example [17, 18].

Turning to the Euler equations (presented here in two dimensions)

$$\begin{aligned} &\text{continuity} && \frac{\partial \rho}{\partial t} + \frac{\partial \rho u}{\partial x} + \frac{\partial \rho v}{\partial y} = 0\,, \\ &\text{x-momentum} && \frac{\partial}{\partial t}(\rho u) + \frac{\partial}{\partial x}(\rho u^2 + p) + \frac{\partial}{\partial y}(\rho u v) = 0\,, \\ &\text{y-momentum} && \frac{\partial}{\partial t}(\rho v) + \frac{\partial}{\partial x}(\rho v u) + \frac{\partial}{\partial y}(\rho v^2 + p) = 0\,, \\ &\text{energy} && \frac{\partial}{\partial t}(\rho e) + \frac{\partial}{\partial x}((\rho e + p)u) + \frac{\partial}{\partial y}((\rho e + p)v) = 0\,, \end{aligned} \tag{5.4}$$

and using

$$\text{set IV} \qquad \begin{aligned} & u' = \frac{u}{\sqrt{\bar{p}/\bar{\rho}}},\ v' = \frac{v}{\sqrt{\bar{p}/\bar{\rho}}},\ \rho' = \frac{\rho}{\bar{\rho}},\ p' = \frac{p}{\bar{p}}, \\ & t' = t\frac{\sqrt{\bar{p}/\bar{\rho}}}{l},\ x' = \frac{x}{l},\ y' = \frac{y}{l},\ e' = \frac{e}{\bar{p}/\bar{\rho}}. \end{aligned} \tag{5.5}$$

to make the equations dimensionless, no dimensionless number appears and the freestream velocity is given by $|\mathbf{v}| = \sqrt{\gamma}\, M$, [19].

Comments on 4:

Well-known are the needs to test flight vehicles like airplanes, spaceplanes and helicopters in wind tunnels as well as ships in watertunnels and the role the dimensionless numbers play in this regard.

But there exists also a great interest to have the potential to investigate the mechanical and thermal behavior of other devices like hydraulic engines (pumps, compressors, turbines, etc.) and edificial structures like buildings, bridges, etc. in ground based facilities. In general this requires to design and fabricate models of these devices, which must be geometrically similar. The question has to be posed under which conditions the physical behavior of the models is similar to the original ones, for example the flow around an airplane. The answer is outlined in very detail in Chap. 8.

References

1. Hirschel, E.H., Weiland, C.: Selected Aerothermodynamic Design Problems of Hypersonic Flight Vehicles. Springer, Berlin/Heidelberg and AIAA (Progress in Astronautics and Aeronautics), Reston USA (2009)
2. Görtler, H.: Dimensionsanalyse. Theorie der physikalischen Dimensionen und Anwendungen. Springer, Berlin/Heidelberg (1975)
3. Hirschel, E.H.: Basics of Aerothermodynamics, 2nd edn. Springer, Berlin/Heidelberg (2015)
4. Krek, R.M., Eitelberg, G., Kastell, D.: Hyperboloid Flare Experiments in the HEG Facility. Deutsche Forschungsanstalt für Luft- und Raumfahrt, DLR-IB 223-95 A **43** (1995)
5. Hirsch, C.: Numerical Computation of Internal and External Flows, vol. I. Wiley, New York, Reprint (1997)
6. Schlichting, H., Gersten, K.: Boundary Layer Theory, 8th edn. Springer, Berlin/Heidelberg (2000)
7. Becker, E.: Gasdynamik. Teubner Verlag, Stuttgart (1966)
8. Zierep, J.: Theoretische Gasdynamik. G. Braun Verlag, Karlsruhe (1972)
9. Liepmann, H.W., Roshko, A.: Elements of Gasdynamics. Wiley, New York (1957)
10. Oswatitsch, K.: Gas Dynamics. Academic Press, New York (1956)
11. Fourier, J.-B.-J.: Théorie analytique de la chaleur. Firmin Didot et Fils (1822)
12. Barenblatt, G.I.: Scaling phenomena in fluid mechanics. Cambridge University Press (1994)
13. Steinrück, H. (ed.): Asymptotic Methods in Fluid Mechanics: Survey and Recent Advances. Springer, Wien/New York (2010)
14. Federmann, A.: Über einige Integrationsmethoden der partiellen Differentialgleichungen erster Ordnung. Annalen des Polytechnischen Instituts Peter der Große zu St. Petersburg (in Russian), Vol. 16, 97–154 (1911)
15. Buckingham, E.: On Physically Similar Systems: Illustration of the Use of Dimensional Equations. Phys. Rev. **4**, 345–376 (1914)
16. Barenblatt, G.I.: Scaling, self-similarity and intermediate asymptotics. Cambridge University Press (1996)
17. Gross, A., Weiland, C.: Numerical Simulation of Separated Cold Gas Nozzle Flows. J. Propulsion Power **20**(3), 509–519 (2004)
18. Gross, A., Weiland, C.: Numerical Simulation of Hot Gas Nozzle Flows. J. Propulsion and Power **20**(5), 879–891 (2004)
19. Weiland, C.: A finite difference method for the calculation of three dimensional supersonic flow fields past blunt bodies. ESA-TT-366 (1977)

Chapter 6
Dimensionless Numbers—Similarity Parameters: A Look at the Name Holders

In Chaps. 2, 3 and 4 we presented various methods with which a count of dimensional numbers, depending on different methods, can be derived. Obviously, Buckingham's Π theorem has, there is no doubt, the capacity with the greatest possible extent. Historically many of the similarity quantities for the first time were formulated as a single event, in particular those known from the 18th and the first part of the 19th century. Therefore, the appearance of the dimensionless numbers obviously was an evolutionary process. Most of the power products, later noted as dimensionless numbers or similarity parameters, were established before the mathematical calculus of the analysis of dimension, Buckingham's Π theorem, was developed.

In this Chapter 20 dimensionless numbers are identified, which bear the name of famous scientists and engineers and which are of major importance in fluid dynamics and heat transport. For these dimensionless numbers the historical background is highlighted with a short biography of the person, whose name the number bears.

The potential of application and a discussion of typical problems, when applied, are stated. For illustration reasons some examples of applicability[1] in order to strengthen the comprehension are presented.

Three dimensionless numbers are listed, which are not connected with a name. There is also one quantity which has the character of a similarity parameter, but possesses a dimension. Finally eleven dimensionless numbers are presented, which are not often used, but illustrate interesting aspects.

[1] In the Chaps. 2, 3 and 4 we distinguish the quantities for dimensional reference values (quantities with a bar $\Longrightarrow \bar{x}$), dimensional variables (clean quantities $\Longrightarrow x$) and dimensionless variables (quantities with a prime $\Longrightarrow x'$). In this chapter we mainly use dimensional reference values ($\bar{x}$). For convenience the bar is always dropped.

C. Weiland, *Mechanics of Flow Similarities*,
https://doi.org/10.1007/978-3-030-42930-0_6

6.1 The Euler Number

Leonard Euler born April 15, 1707 in Basel, Switzerland and death in St. Petersburg, Russia on Sept. 18, 1783 was a mathematician, physicist and astronomer. He went 1727 to the Imperial Russian Academy of Sciences in Saint Petersburg upon a recommendation of Daniel Bernoulli and filled a position in the mathematical department. Due to the turmoil situation in Russia Euler left St. Petersburg on 1741 in order to take over a post at the Berlin Academy, where he stayed for 25 years. In 1766 he again accepted an invitation to the St. Petersburg Academy, mainly driven by the increasingly shattered relationship with Frederick the Great of Prussia. He remained there until his death in 1783. Leonard Euler was an extremely productive scientist. He wrote hundreds of scientific articles (approximately 900) and more than 20 books on a large variety of mathematical and physical problems, [1, 2]. He was the first who had formulated the hydromechanical differential equations in three dimensions. We know these equations as the convective part (including the pressure) respectively the inviscid part of the fundamental fluid dynamic momentum equations (Eq. (4.7)), today called the Euler equations. Further he had given important contributions to many other topics.[2] One of the dimensionless numbers bears his famous name, [3, 4].

Leonard Euler

The Euler number is the ratio of the pressure force to the force of inertia, Eqs. (2.18), (3.2), (4.15).

$$Eu = \frac{p}{\rho u^2} \quad \text{or} \quad Eu = \frac{\Delta p}{\rho u^2} \, .$$

Often we meet the Euler number in different forms. For example for an ideal gas we obtain

$$Eu = \frac{1}{\gamma M^2} \, , \tag{6.1}$$

[2]Contributions to mathematics: geometry, trigonometry, graph theory, number theory, topology, infinitesimal calculus, algebra. Contributions to physics: continuum mechanics (fluid dynamics), optics, lunar theory, stability theory, ballistics, equation of gyroscope.

whereby also the Mach number and the ratio of specific heats are defined as dimensionless numbers with $M = u/a$ and $\gamma = c_p/c_v$. Another form of the Euler number is given by the definition of the pressure coefficient.

$$c_p = \frac{\Delta p}{\frac{1}{2}\rho u^2} = \frac{\Delta p}{q} \qquad q \text{ the dynamic pressure .}$$

6.1.1 A Characteristic Application of the Euler Number

A model of a car is investigated in a water tunnel, Fig. 6.1. In the table below the quantities of the model (M), the real car (F) and the flow conditions are listed (Table 6.1).

The maximum Reynolds number for the model has the value

$$Re_M^{max} = \frac{u_M^{max} l_M}{\nu_M} = 2 \cdot 10^6$$

Since similar flow is considered, the Reynolds number has to be the same and we obtain

Fig. 6.1 Sketch of car motion in water tunnel u_M^{max} and in air u_F^{max}

Table 6.1 List of quantities regarding the Euler number example

	Model in water	Real car in air
Length l	0.4 m	4 m
u_{max}	$5 \ \frac{\text{m}}{\text{s}}$	$8 \ \frac{\text{m}}{\text{s}}$
Kinematic viscosity ν	$10^{-6} \ \frac{\text{m}^2}{\text{s}}$	$16 \cdot 10^{-6} \ \frac{\text{m}^2}{\text{s}}$
Density ρ	$1000 \ \frac{\text{kg}}{\text{m}^3}$	$1.2 \ \frac{\text{kg}}{\text{m}^3}$

$$Re_M^{max} = Re_F^{max} \qquad \Longrightarrow \qquad u_F^{max} = 8\frac{\mathrm{m}}{\mathrm{s}} \ .$$

Further when we have a view at identical Euler numbers ($Eu_M = Eu_F$) we find

$$\frac{\Delta p_M}{\rho_M (u_M^{max})^2} = \frac{\Delta p_F}{\rho_F (u_F^{max})^2} \qquad \Longrightarrow \qquad \frac{\Delta p_F}{\Delta p_M} = 0.003072 \ ,$$

where similar flow means that the dynamic pressure (stagnation pressure) of the model in the water tunnel (Δp_M) is by the factor 1./0.003072 higher than that of the real car driving through air (Δp_F).
Note: A simple evaluation of Bernoulli's equation ($p_0 = p_\infty + 1/2\rho_\infty u_\infty^2$) yields the same result

$$\Delta p_M = p_M - p_\infty = 0.125 \cdot 10^5 \ ,$$

$$\Delta p_F = p_F - p_\infty = 0.000384 \cdot 10^5 \ .$$

6.2 The Froude Number

William Froude was an English engineer for ship design and hydrodynamics. He was born on Nov. 28, 1810 in Darlington, England and died on May 4, 1879 in Simontown, South Africa, while on holiday. William Froude had worked intensively in the field of ship design and was the first who had 1872 determined the drag of a ship. He did it by establishing a formula (now known as the Froude number) by which the results of a ship model test (small-scale test) was used to predict the behavior (drag and ship hull) of a full-sized ship, [5]. The number, which bears his name, describes the ratio between the force of inertia and the force of gravitation. Moritz Weber, [6], was the scientist, who first had called the above relation the Froude number Fr. The number is further important for the development of water waves due to ship movements, for shallow water as well as flume flows, [7]. W. Froude was elected a Fellow of the Royal Society in 1870 and in 1876 he received the Royal Medal from the Royal Society.

William Froude

By Eqs. (2.16), (3.2), (4.15) the Froude number is defined as

$$Fr = \frac{u^2}{gl} \,, \tag{6.2}$$

with u being the velocity, g the gravitational acceleration and l a reference length. It presents the ratio between the force of inertia and the gravitational force. Often and in particular for ship design and ship flows the definition

$$Fr' = \frac{u}{\sqrt{gl}} = \sqrt{Fr}$$

is used.

6.2.1 Aspects of Ship Design Due to the Production of Water Waves in the Light of the Froude Number

Moving ships create water waves (bow waves, stern waves, etc.) and hence wave drag. This drag causes costs and reduces the energy efficiency of the ship. Therefore the ship designer has to construct a ship, which produces a system of water waves, which is as energy-saving as possible. The front and the rear part of a moving ship induce strong waves, whereas along the forward and rearward shoulder of the ship also water waves are produced. But of lower strength. The main task is now to find a ship length (for a given velocity), where the forward wave (bow wave) interferes with the rearward wave (stern wave) in the sense that wave cancellation occurs (wave trough encounters crest of wave), [8].

The wave length[3] λ produced by the moving ship (or a moving pressure point) in a deep water approximately is

$$\lambda = 2\pi \frac{u^2}{g} = 2\pi Fr \cdot l \,,$$

$$u = \lambda f = \lambda \frac{1}{T} \,, \qquad T = \sqrt{\frac{2\pi\lambda}{g}} \,,$$

with u being the speed of the wave, T the period of the wave, f the frequency of the wave and l length of the ship. Experience has shown that the most efficient ship

[3]Note: The wave length is that of the cross wave. For definition see [8, 9].

lengths are given by $1/(2\pi Fr) = l/\lambda = 0.5,\ 1.5,\ 2.5,\ 3.5, \ldots$ (uneven multiples of $\lambda/2$). Therefore favorable Froude numbers are

$\frac{l}{\lambda}$	Fr'
0.5	0.564
1.5	0.325
2.5	0.252
3.5	0.213
⋮	⋮

Unfavorable Froude numbers are $1/(2\pi Fr') = l/\lambda = 1.0,\ 2.0,\ 3.0, \ldots$ (even multiples of $\lambda/2$), namely

$\frac{l}{\lambda}$	Fr'
1.0	0.399
2.0	0.282
3.0	0.230
⋮	⋮

Figure 6.2 indicates left the wave image where the ship length is an uneven multiple of $\lambda/2$ ($l/\lambda = 2.5 \implies Fr' = 0.252$). The right part of the figure exhibits the same for an even multiple of $\lambda/2$ ($l/\lambda = 2.0 \implies Fr' = 0.282$). One can clearly discern that the wave structure is much more distinct in the right part compared to the left part, where obviously in the wake of the ship a more or less complete wave cancellation has taken place.

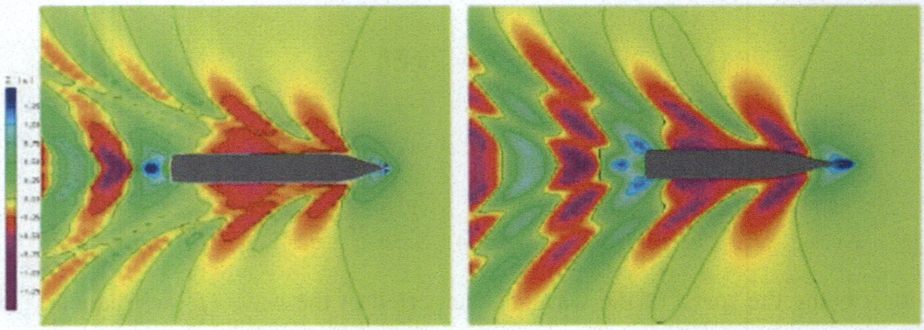

Fig. 6.2 Different wave images of a moving ship. Left: Froude number $Fr' = 0.252$; right: Froude number $Fr' = 0, 282$. Figure taken from [8]

Let two ships be geometrically similar. Then we can write for the ship length ratio

$$k = \frac{l_{ship}}{l_{model}} ,$$

for surfaces A and masses M it follows

$$k^2 = \frac{A_{ship}}{A_{model}} \quad \text{and} \quad k^3 = \frac{M_{ship}}{M_{model}} ,$$

and with the gravitational acceleration g, which is equal for ship and model we obtain

$$\sqrt{k} = \frac{u_{ship}}{u_{model}} .$$

This is the parametrical environment for model experiments.

The flows past the ship and the model are similar, when both have the same Froude number $Fr' = u/\sqrt{gl}$. This is true also for the wave structure and the wave drag D. Therefore we have

$$D_{model} = \frac{D_{ship}}{k^3} . \tag{6.3}$$

Given are the length of a ship $l_{ship} = 200$ m and the length of a model $l_{model} = 5$ m which means $k = 40$. With relation Eq. (6.3) we recognize that the wave drag of the ship is 64000 higher than that of the model.

6.3 The Weber Number

Moritz Gustav Weber was a German engineer. Born in Leibzig, Germany, on Juli 18, 1871 he raised up in Hannover and died in Neuendettelsau, Germany on June 10, 1951. In 1904 he became a professor for mechanics at the Technical University of Hannover. Then in 1913 he was appointed as professor for ship design at the Technical University in Berlin-Charlottenburg, where he remained until his retirement in 1936. Moritz Weber had worked intensively in the field of the mechanics of similarity and the analysis of dimensions, especially with applications to ship design and ship flows. His opinion was, that these disciplines had much more significance for technical physics, than to be only a guidance for the execution and evaluation of model tests, [6, 10]. The similarity number, which carries his name, Weber number We, characterizes the ratio of inertial force to surface tension force.

Moritz Gustav Weber

By Eq. (3.2) the Weber number is defined as

$$We = \frac{\rho u^2 l}{\sigma}, \tag{6.4}$$

and depicts the rate of force of inertia to surface tension force or capillary force.

6.3.1 An Application of the Weber Number for the Evolution of a Diesel Injection Jet

The Weber number is often used for the description of the processes of spray generation during the fuel injection procedure of gasoline or Diesel motors.

Figure 6.3 shows a sketch of a part of a typical Diesel fuel injection system. When the fuel leaves the nozzle, the jet sustains at first the so called primary decay, where the jet becomes wavy and forms first holes. These holes scarify further and discrete ligaments come into play. The primary decay phase is finished, when the ligaments collapse to coarse drop-shaped structures, [11].

Then the secondary decay procedure starts with a further decomposition of the large drops and the remaining ligaments into small drops. This happens due to the momentum transfer between the gas (air) and the fluid phase of the spray. Shear forces and vaporization care for further reduction of the drop size. The inertial force conveys the decay of the drops, while the surface tension tries to prevent it. This process is described by the Weber number

$$We = \frac{\rho u_{rel}^2 d}{\sigma},$$

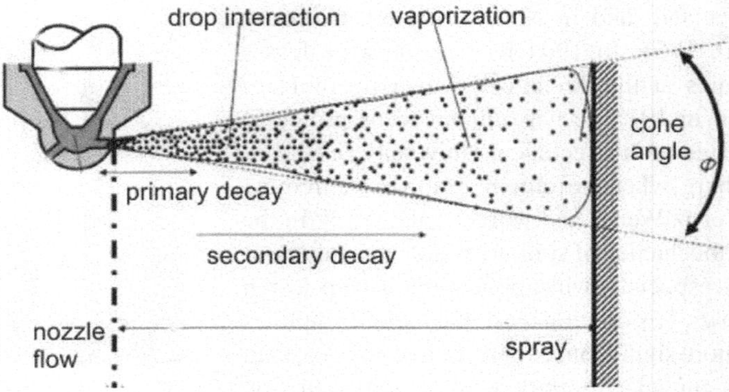

Fig. 6.3 Sketch of a typical Diesel fuel injection system. Figure taken from [11]

with ρ being the density of the drops, d the diameter of the drops, u_{rel} the relative speed between ambient gas (air) and drops and σ the surface tension of the drops, [12–14].

There exists a large number of various forms of decay processes of drops, which are described by different regimes of Weber numbers. However for a main decomposition process a critical Weber number can be defined by

$$We_{crit} \approx 12 ,$$

which implies that for higher Weber numbers the drops collapse (The drop-stabilizing of surface tension are too small compared to the destabilizing inertial forces.), and in contrary for lower Weber numbers the drops (and their formation) are stable, [12, 15].

6.4 The Grashof Number

Franz Grashof was a German engineer. He was born on Juli 11, 1826 in Düsseldorf and died on Oct. 26, 1893 in Karlsruhe, Germany. In 1863 be became a professor for the theory of practical mechanics at the Polytechnic Institute (later Technical University) in Karlsruhe, where he remained until 1891. Several times he acted as the general director of this institution. Franz Grashof was a founder member of the VDI—"Verein Deutscher Ingenieure[4]"—in 1856 and was its first director before he retired from this post in 1890. He has worked in the fields of theoretical-mechanical engineering, mechanics of materials, hydraulics and thermodynamics. He gave lectures in analytical mechanics, elasticity and stability, hydraulics and practical mechanics, [16]. In all these fields he published textbooks. The dimensionless fluid dynamic number denoting the ratio of lift force to inertial force times the square of the Reynolds number bears his name Grashof number[5] Gr, which is used frequently for the description of free convection flows.

Franz Grashof

[4] Society of German Engineers.

[5] Sometimes also described as the ratio of lift force to friction force.

In Eq. (3.2) we have formulated the relation $-\frac{gl}{u^2} \cdot \frac{\Delta\rho}{\rho}$, where $\frac{\Delta\rho}{\rho}$ is defined by $\frac{\rho_w - \rho_\infty}{\rho_\infty} = -\beta(T_w - T_\infty)$, with $\beta = -\frac{1}{\rho}\frac{\partial\rho}{\partial T}$ being the volumetric coefficient of thermal expansion, T_w the wall temperature and T_∞ an appropriate reference temperature. We find then a relation which includes the Grashof number and the Reynolds number

$$\frac{gl\beta}{u^2} \cdot (T_w - T_\infty) = \underbrace{\left(\frac{\mu}{\rho u l}\right)^2}_{1/Re^2} \cdot \underbrace{\frac{\rho^2 g\beta l^3}{\mu^2} \cdot (T_w - T_\infty)}_{Gr} .$$

For a perfect gas we evaluate the relation $\beta = -\frac{1}{(p/RT)}\frac{\partial(p/RT)}{\partial T} = \frac{1}{T}$ (Boussinesq approximation) and obtain

$$\frac{gl}{u^2} \cdot \frac{T_w - T_\infty}{T_\infty} = \left(\frac{\mu}{\rho u l}\right)^2 \cdot \frac{\rho^2 g l^3}{\mu^2} \cdot \frac{T_w - T_\infty}{T_\infty} = \frac{1}{Re^2} \cdot Gr .$$

6.4.1 The Importance of the Grashof Number for Free Convection Flows

We consider the free convection flow which is generated along a heated vertical wall, Fig. 6.4. The flow creates a viscous and a thermal boundary layer, whereby the thickness of the viscous one is denoted by δ and the thermal one by δ_t. Further the velocity profile and the temperature profile are drawn in this figure, [17].

Let the wall temperature be $T_w = 60\,°\mathrm{C} = 333\,\mathrm{K}$, the temperature of the ambient air $T_\infty = 20\,°\mathrm{C} = 293\,\mathrm{K}$, the kinematic viscosity $\mu/\rho = \nu = 16.92 \cdot 10^{-6}\,\mathrm{m^2/s}$ and the vertical length $L = 0.6\,\mathrm{m}$, then the Grashof number yields

$$Gr = \frac{gL^3}{\nu^2} \cdot \frac{T_w - T_\infty}{T_\infty} = 1.01 \cdot 10^9 \qquad (6.5)$$

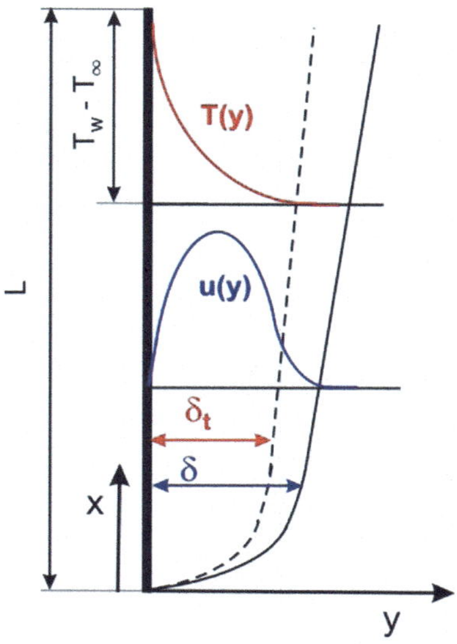

Fig. 6.4 Sketch of a free convection flow situation along a vertical heated wall

6.5 The Reynolds Number

Osborne Reynolds was a British mathematician and physicist. On August 23, 1842 he was born in Belfast, Ireland and died on Feb. 21, 1912 in Watchet, England. He studied mathematics at Queens' College in Cambridge and graduated in 1867. In the following year (1868) he applied for and was elected to the newly installed chair of engineering at the Victoria University of Manchester, where he remained as professor of engineering until 1905, [19, 20]. His scientific interests were versatile. Besides his well-known contributions to fluid dynamic problems, he published papers on the physics of gases, liquids and granular materials as well as on solar and cometary matters. In 1883 he published his famous paper regarding the fundamental experimental observations with respect to laminar and turbulent flows and their dependency on the dimensionless quantity ul/ν, [21], see also [22]. A. Sommerfeld has proposed in 1908 to call this quantity in honor to him Reynolds number $\left(Re = \dfrac{ul}{\nu}\right)$.

Osborne Reynolds

Fig. 6.5 Original test rig of O. Reynolds' experiment, [21]

The Fig. 6.5 exhibits a sketch of O. Reynolds' original apparatus for the color filament experiment.

With the formal methods presented in the Chaps. 2, 3 and 4 the dimensionless quantity known as the Reynolds number Re could be derived. The equations Eqs. (2.12), (3.2), (4.15), (4.18) indicate this

$$Re = \frac{\rho u l}{\mu} = \frac{ul}{\nu},$$

with ρ being the density, u the velocity, l a characteristic length, μ the dynamic viscosity and ν the kinematic viscosity. This outstanding quantity is applied intensively in wind tunnel techniques and describes some of the aspects of laminar and turbulent flows as well as transitional flows. The original definition of the Reynolds number is given by the ratio of inertial force to friction force. But it has been turned out, that this definition fails in some specific flow cases.

When we have a look on the rate of inertial force F_i, Eq. (3.1b), we recognize the proportionality $\frac{\partial}{\partial x}(\rho u^2) \Longrightarrow \frac{\rho u^2}{l} \sim F_i$. In the case of a fully developed pipe flow, Fig. 6.6, there is no change of the velocity u in x-direction, which means $\frac{\partial}{\partial x}(\rho u^2) = 0 = F_i$. This results in $Re = 0$, which is obviously wrong. Therefore the definition of the Reynolds number as ratio of inertial force to friction force fails in this example. K. Oswatitsch has pointed to this problem in 1963, [23]. He proposed to define the

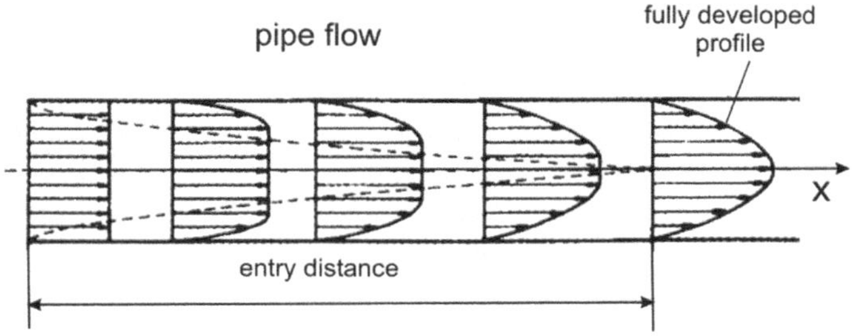

Fig. 6.6 Sketch of the flow profile development in a pipe during the entry phase

Reynolds number as the ratio of momentum flux to shear stress. Hirschel [24], has defined it as convective momentum transfer to molecular momentum transfer, which seems to be the most general and most comprehensive definition.

The correctness of the advanced definition, given above, can be demonstrated by the following example found in [4]. Given the Couette flow, Fig. 6.7, where the velocity is defined by $u = \frac{U}{d}y$. Then we can calculate the momentum flux J by

$$J = \frac{\rho}{d}\int_0^d u^2 dy = \frac{\rho\,U^2}{d^3}\int_0^d y^2 dy = \frac{\rho\,U^2}{3} \sim \rho\,U^2\,, \tag{6.6}$$

the shear stress is

$$\tau = \mu\frac{\partial u}{\partial y} = \mu\frac{U}{d}\,,$$

and it follows

$$Re = \frac{J}{\tau} = \frac{\rho\,U d}{\mu}\,.$$

Obviously Eq. (6.6) is more appropriate for the definition of the Re number than Eq. (3.1b).

There are some other interpretations of the Reynolds number. Two of them are described below.

- The Reynolds number reflects the ratio of the kinetic energy E_{kin} (twice) of a volume V, which is moved by a velocity u and the friction work E_f, which is performed, when this volume V is shifted along the length l.

$$Re = \frac{\rho u l}{\mu} = \frac{\rho u^2 l^3}{\mu u l^2} = \frac{\rho V u^2}{\mu u A} = \frac{2E_{kin}}{E_f}\,.$$

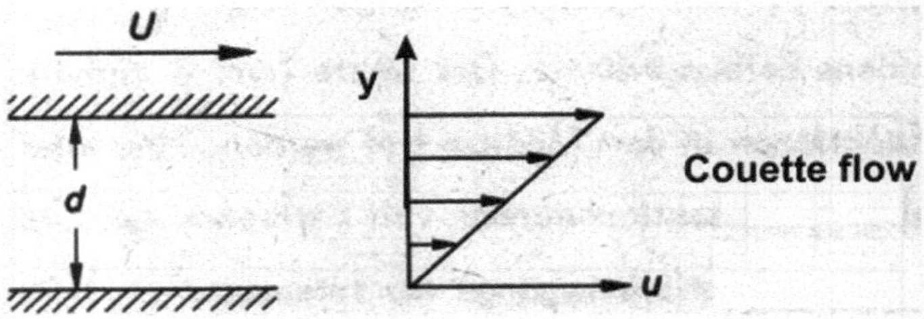

Fig. 6.7 Sketch of Couette flow

- The friction force of a sphere moved in a fluid is constituted corresponding to Stokes' law by $F_f = 6\pi\mu r u$. When this sphere is shifted along the length l with $l = 2r$, the energy due to the friction force amounts to $E_f = F_f \cdot l = 3\pi\mu u l^2$. The kinetic energy of the moved sphere is $E_{kin} = \frac{1}{2}mu^2 = \frac{1}{12}\pi\rho u^2 l^3$ and it follows

$$\frac{E_{kin}}{E_f} = \frac{\pi\rho u^2 l^3}{12 \cdot 3\pi\mu u l^2} \sim \frac{\rho u l}{\mu} = Re \ .$$

6.5.1 The Importance of the Reynolds Number: Some Characteristic Examples

The aerodynamics and aerothermodynamics of flight vehicles like military and civil aircraft, missiles for defence functions as well as space vehicles like capsules as well as winged and lifting reentry systems is one of the major fields where the Reynolds number plays an essential role. This is true in particular for flow field classifications and the assessment of the validity of wind tunnel testing. As already mentioned earlier, it is often very difficult in experiments or test facilities to meet all the relevant similarity parameters. The main similarity parameters for classical wind tunnel tests of flight vehicles are the Mach number and the Reynolds number. Most of the wind tunnels are able to comply with the Mach number, but often fail in the reproduction of the Reynolds number by one or two orders of magnitude. To overcome this problem, in particular for the transonic Mach number range, in Europe a huge cryogenic wind tunnel was built, the European Transonic Wind Tunnel ETW, located in Cologne, Germany, Fig. 6.8, [25].

Fig. 6.8 European Transonic Wind Tunnel ETW in Cologne, Germany, [25]

Table 6.2 ETW's test conditions

Test section size	2.00 m × 2.40 m length 9.00 m
Mach number range	0.15 to 1.35
Reynolds number	up to $8.5 \cdot 10^7$ (semispan model)
Test section pressure range	1.25 to 4.5 bar
Temperature range of test gas	110 –313 K
Typical span width of model	1.6 m
Test gas	Nitrogen

The high Reynolds numbers in the ETW are enabled by the reduction of the test gas temperature and the raise of the gas pressure. This cumulates to a factor of approximately six compared to facilities with atmospheric conditions. The tunnel allows to independently duplicate Mach number, Reynolds number and dynamic pressure. All the test conditions are listed in Table 6.2.

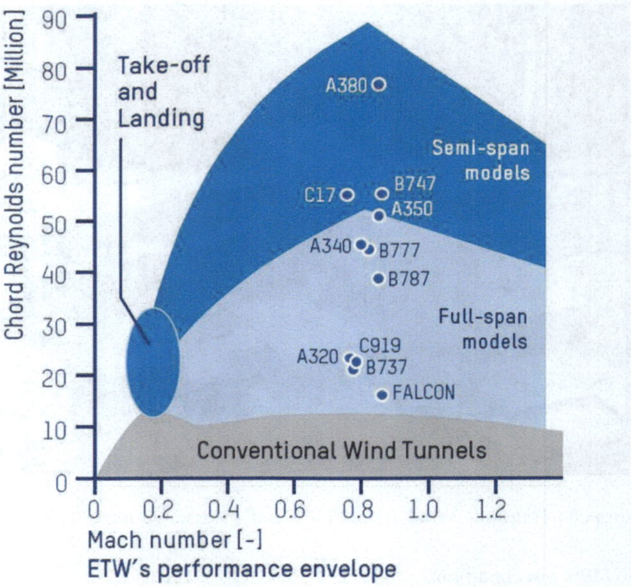

Fig. 6.9 Flight envelope of European Transonic Wind Tunnel ETW, [25]

Fig. 6.10 European Transonic Wind Tunnel ETW. View in test section with a civil aircraft model (left), model cooling chamber (right), [25]

As the performance envelope exhibits, Fig. 6.9, all present-day civil aircraft,—including the huge A380-, can be tested in the ETW, where the free-flight values for the Mach number **and** the Reynolds number are completely duplicated. Figure 6.10 gives an insight into different parts of the ETW facility.

The following figures show the eminent importance of the Reynolds number for the flow structures.

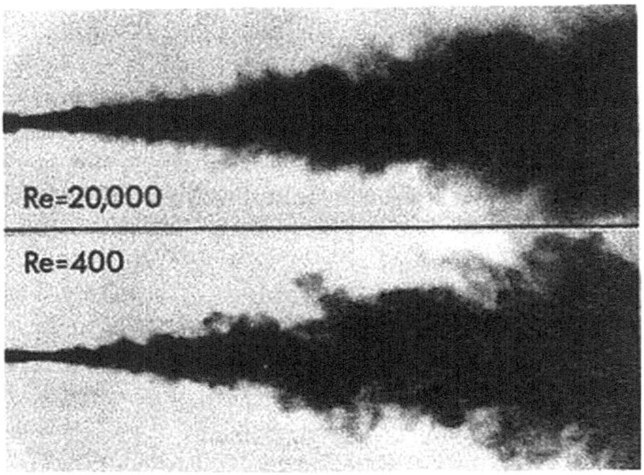

Fig. 6.11 Jet flows of different viscosity in ambient water, [26]

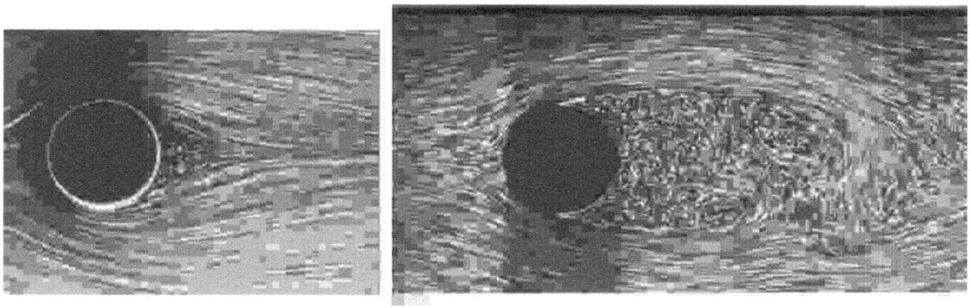

Fig. 6.12 Flow around a circular cylinder with different Reynolds numbers. Left photograph by Sadatoshi Taneda ($Re = 9.6$), right photograph by Werlé & Gallon ($Re = 2000$), [27]

Figure 6.11 shows jet flows in ambient water. The viscosity of the jet in the upper figure is low, leading to a high Reynolds number ($Re = 20000$) and high in the jet of the lower figure, where the Reynolds number amounts to $Re = 400$. As one can observe the eddies are large for $Re = 400$ and small for $Re = 20000$, [26].

In Fig. 6.12 a circular cylinder is moved through a water tank. The flow presented in the left part of the figure has the Reynolds number $Re = 9.6$, where in the aft part of the cylinder the flow has separated and forms a pair of recirculating eddies. In the right part of the figure the flow has a Reynolds number $Re = 2000$. The laminar boundary layer over the front separates and breaks up into a turbulent wake, [27].

6.6 The Péclet Number

Jean Claude Eugéne Pèclet was a French scientist. Born at Besancon on Feb. 10, 1793 and dying in Paris on Dec. 8, 1857. He was one of the first students of the École Normale in Paris with Gay-Lussac being his teacher. In 1816, he became an assistant professor of physics at the College of Marseille, where he stayed until 1827. Then he returned to Paris and was a founder and the first director of the Central School of Arts & Manufacture in 1829. He retired from this post in 1852, [28]. Pèclet conducted studies in ventilation, fresh air requirements and removal of water vapor and odours. His experience seems to have mainly been in hot-air heating installations. Pèclet has installed his air heating systems in many important buildings between 1840 and 1854, [29]. One of these was the chamber of deputies in Paris, 1841. Besides others he wrote a book about heat, published in 1830.

Jean Claude Eugéne Pèclet

The Péclet number has the same structure as the Reynolds number. By Eqs. (3.9) and (4.19) its definition is

$$Pe = \frac{lu}{k} ,$$

with u being the velocity, k the thermal diffusivity and l a characteristic length.

The Péclet number embodies the ratio of the heat convection to the heat conduction (diffusive energy transfer). The analogy[6] with the Reynolds number can be expressed by

$$Pe = \frac{lu}{k} = \frac{lu}{\nu} \cdot \frac{\nu}{k} = Re \cdot Pr ,$$

with $Pr = \nu / k$ being the Prandtl number.

Péclet numbers around unity ($Pe \approx 1$) means that heat convection and heat conduction are more or less of the same order of magnitude. Heat is transferred mainly by conduction in flows with small Péclet numbers ($Pe << 1$), and by convection when the Peclet number is large ($Pe >> 1$).[7]

[6] The transport coefficient of the momentum is replaced by the transport coefficient of the energy.

[7] There exists an analogue Péclet number with respect to mass transport, namely $Pe' = \frac{lu}{D_{ij}}$ with D_{ij} the binary diffusivity.

6.6.1 An Example from the Geology (Geodynamics)

A descriptive application of the Péclet number is given by the following example. The thermal process in the Earth' lithosphere is considered. We remember, that $Pe < 1$ means that the heat conduction (diffusion) dominates, and when $Pe > 1$ convection dominates. For $Pe \approx 1$ the influence of both is of the same order. Assuming that the deformation in most mountainous regions involves only Earth crust, Fig. 6.13, which has a typical thickness l of approx. 35000 m. The question what the uplift velocity u is in an active mountain region above which convection is more dominant than conduction (diffusion) can be answered by the Peclet number, [30, 31]:

$$u = \frac{Pe\,k}{l}\,, \tag{6.7}$$

with the typical value for the thermal diffusivity of crustal rocks $k = 10^{-6}\ \mathrm{m^2/s}$ we obtain

$$u \geqq \frac{1 \cdot 10^{-6}}{3.5 \cdot 10^{4}} = 2.86 \cdot 10^{-11}\ \frac{\mathrm{m}}{\mathrm{s}} \quad \text{or} \quad u \gtrapprox 0.9\ \frac{\mathrm{mm}}{year}\,. \tag{6.8}$$

Fig. 6.13 Deformation in mountainous regions

6.7 The Eckert Number

Ernst Rudolph Georg Eckert

Ernst Rudolph Georg Eckert was born in Prague, now the Czech Republic, on Sep. 13, 1904 and died on July 8, 2004 in St. Paul, Minnesota, USA. He had his studies at the Technical University at Prague, where he graduated in 1927 and where he prepared his doctoral thesis in 1931. His scientific work has encompassed all aspects of heat and mass transfer (convection, conduction, radiation, friction, etc.), [32]. His first employment was at the German Research Institute for Aerospace (DFL) in Braunschweig. After World War II he worked some years in the jet propulsion area of the Wright-Paterson Air Force Base in Dayton, Ohio, before be became in 1951 a professor of mechanical engineering at the University of Minnesota, where he remained until his retirement in 1973. One of his main interests was the thermal management of gas turbines. Other areas of his work were thermal radiation, duct- and tube-flow heat transfer, natural convection, high speed flow heat transfer (the problem of re-entry flight from space) and film cooling along engine walls, [33]. He published a hugh number (550) of papers and books.

With Eq. (3.9) we have shown, that the Eckert number can be derived by the evaluation of the ratio of the heat flux due to friction to the convective heat flux,

$$\frac{\mu u^2}{l^2} \cdot \frac{l}{\rho c_p T u} = \frac{u^2}{c_p T} \cdot \frac{\mu}{\rho u l} = Ec \cdot \frac{1}{Re} .$$

For perfect gas the Eckert number can be written in the form

$$Ec = \frac{u_\infty^2}{c_{p,\infty} T_\infty} = (\gamma - 1) M_\infty^2 , \quad (6.9)$$

where for the freestream conditions we have u_∞ the velocity, $c_{p,\infty}$ the specific heat at constant pressure, T_∞ the temperature and M_∞ the Mach number.

Often the Eckert number is also defined by a temperature difference ΔT, e.g. the wall temperature T_w and a reference temperature T_∞ along the wall ($\Delta T = T_w - T_\infty$).

$$\tilde{Ec} = \frac{u^2}{c_p \Delta T} . \quad (6.10)$$

There are several interpretations of the Eckert number:

- the Eckert number is a measure of the influence of the dissipation in a flow,

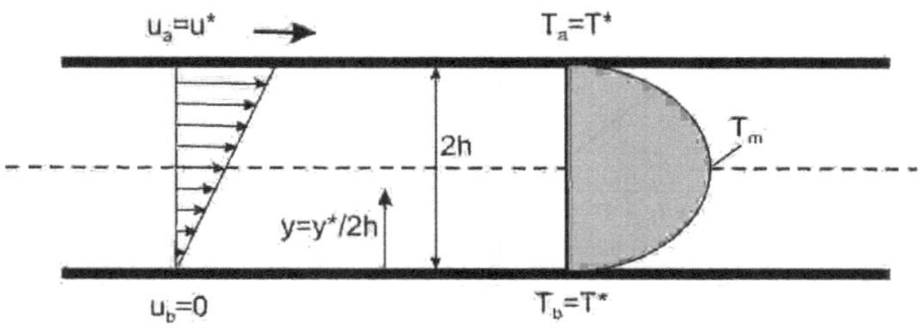

Fig. 6.14 Couette flow: influence of the dissipation on the temperature profile

- the Eckert number is the ratio of kinetic energy to the enthalpy difference between wall and fluid,
- the Eckert number measures the temperature increase from adiabatic compression, [34].

6.7.1 Determination of the Temperature Profile in a Couette Flow

We ask for the temperature profile due to dissipation in a Couette flow, Fig. 6.14. The following equation describes this[8]

$$T - T^* = \frac{1}{2} Pr \cdot \frac{u^2}{c_p}\, y(1-y)\,,$$
$$= \frac{1}{2} Pr \cdot Ec\; y(1-y)\, T^*\,. \qquad (6.11)$$

When we consider an oil flow with $T_a = T_b = T^* = 293$ K, $p = 1$ bar, $Pr = 10400$, $c_p = 1900$ J/kgK we obtain for the two velocities $u^* = 5\,\text{m/s}$ and $u^* = 10\,\text{m/s}$ the Eckert number Ec and the temperature difference in midflow $\Delta T_m = T_m - T^*$

$$u^* = 5\,\text{m/s},\quad Ec = 4.5 \cdot 10^{-5},\quad \Delta T_m = T_m - T^* = 17\,°\text{C}\,,$$
$$u^* = 10\,\text{m/s},\; Ec = 17.9 \cdot 10^{-5},\; \Delta T_m = T_m - T^* = 68\,°\text{C}\,,$$

which seems to be significant temperature increases, (example taken from [7]). In Fig. 6.15 the temperature distribution ΔT due to dissipation by evaluation of Eq. (6.11) for three velocities is given.

---

[8]This equation can be derived from the energy equation, where only the heat conduction and the dissipation play a role (by integration of $\frac{\partial}{\partial y}\lambda\frac{\partial T}{\partial y}$), and when the boundary conditions of the Couette flow are applied, see Fig. 6.14.

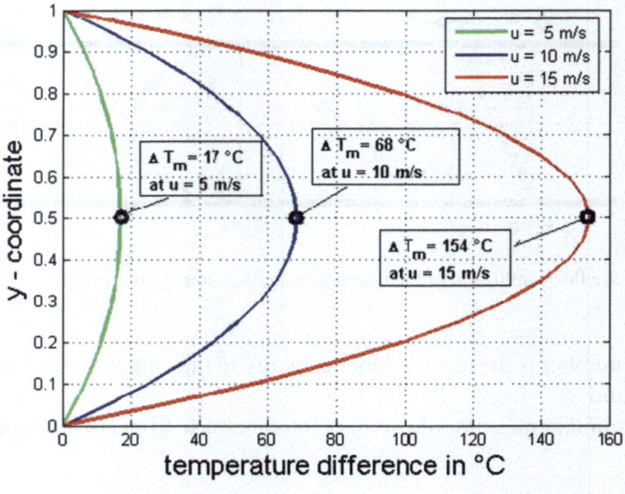

Fig. 6.15 Couette flow: temperature distributions ΔT due to dissipation for three different velocities

6.8 The Damköhler Numbers

Gerhard Damköhler was a German scientist. He was born on March 16, 1908 in Klingenmünster, Germany and died on March 30, 1944 in Braunschweig, Germany.[9] At the University of Munich, Germany, he studied chemistry and prepared his doctoral thesis with K. Fajans. He worked in the fields of chemical engineering, fluid dynamics and thermodynamics. In 1934 Arnold Eucken, at that time the director of the Institute of Physical Chemistry at the University of Göttingen, offered him a position as an assistant, where he then was involved in the development of analytical methods for technical problems in chemical engineering, [35, 36]. Due to a couple of very outstanding publications he received in 1937 the title *Dr. phil. habil.* from the University of Göttingen, a special German form for extended doctoral thesis work. Later in 1937 he accepted a post at the Institute of Propulsion (Motorenforschungsanstalt) of the Aeronautical Research Establishment (Luftfahrtforschunganstalt) in Braunschweig, Germany, at that time headed by Ernst Schmidt, see Sect. 6.12. Two dimensionless numbers relevant in fluid dynamics bear his name.[10]

Gerhard Damköhler

[9]Due to tragic circumstances he obviously took his life.

[10]The major field of application of the Damköhler numbers consists in the discipline of chemical engineering (chemical reactor technology).

Space vehicles experience hypersonic flight conditions during ascent into Earth' orbit and later during their re-entry in Earth' atmosphere. Such space vehicles can be capsules like Apollo and Soyuz or winged configurations like the Space Shuttle Orbiter. In an inertial system the flight speed, for example in a circular 400 km orbit, amounts to $\approx$7670 m/s. During re-entry this speed leads to flight Mach numbers of about 30 with the consequence that the air flow around such vehicles is captured in a thermodynamically non-equilibrium state. The knowledge about the aerodynamics or better the aerothermodynamics is vital for a safe and reliable re-entry stage, [37–39].

For the characterisation of such flows the ratio between the residence time of a fluid particle in a flow region t_{res} and the characteristic reaction time τ, which is needed to have the necessary number of collisions to induce a thermodynamic or chemical process (vibrational excitation, dissociation, chemical reaction, ionisation), is of substantial importance. This ratio is named the first Damköhler number, [24]:

$$DAM1 = \frac{t_{res}}{\tau} \ . \tag{6.12}$$

The following cases can be distinguished:

DAM1 $\Longrightarrow \infty$ the residence time is much larger than the reaction time: $t_{res} >> \tau$; nearly all the thermodynamic and chemical reactions have taken place during the time interval (t_{res}) while the related group of fluid particles passes the flight configuration. This can also be expressed by a reference length. Given a reference length of the configuration l_{ref} we have $l_{ref} = t_{res} v_{ref}$ with v_{ref} a reference velocity of the spacecraft. In that case the air is in thermodynamic and chemical equilibrium.

DAM1 $\Longrightarrow 0$ the residence time is much shorter than the reaction time: $t_{res} << \tau$; no thermodynamic and chemical reactions take place during the pass of the fluid elements along the configuration (illustrated by the reference length l_{ref}). The flow is in a "frozen" state.

DAM1 $\Longrightarrow O(1)$ the residence time is of the order of the reaction time: $t_{res} \approx \tau$; the flow gas is in non-equilibrium, which means that thermodynamic and chemical relaxation processes take place during the pass of the fluid elements along the configuration.

The second Damköhler number is given by Eq. (3.9)

$$DAM2 = \frac{Q}{c_p \Delta T} \ .$$

It compares the process energy of thermo-chemical reactions (e.g.: Energy involved in non-equilibrium real gas effects.) with a related enthalpy difference. $DAM2 \Longrightarrow 0$ means, that the energy given by thermo-chemical reactions can be neglected in the energy balance of the flow.

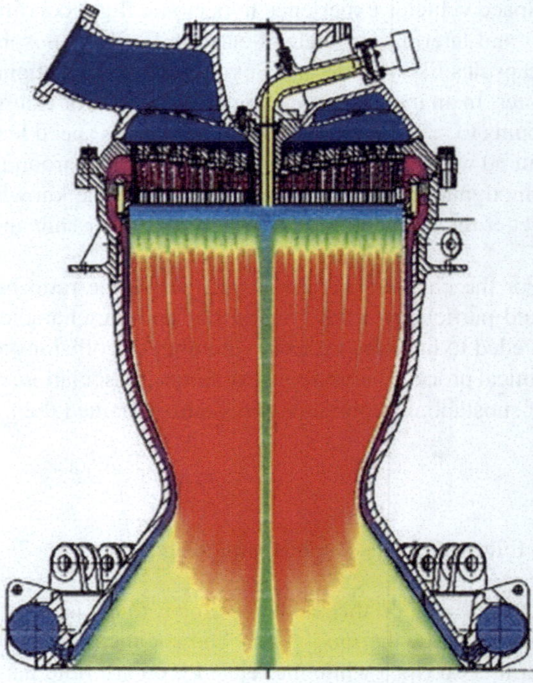

Fig. 6.16 Combustion chamber flow in a rocket motor including throat and a short part of the expansion nozzle, [40]

6.8.1 *Aspects of High-Speed Flows in the Light of the First Damköhler Number (DAM1)*

The first Damköhler number has its importance for:

- hypersonic flows, where the gas is in a thermo-chemical non-equilibrium state.
- combustion chambers of rocket motors, where the degree of completed reactions (e.g.: $2H_2 + O_2 \Longrightarrow 2H_2O$ in a liquid hydrogen-oxygen rocket engine) plays an important role for the cooling concept of the combustion chamber and the nozzle walls as well as for the determination of the thrust.

Figure 6.16 shows the reaction process in the combustion chamber of the Vulcain2 motor, which is the main engine of the Ariane V ESC-A/B rocket, [40]. Behind the throat, when the gas is strongly expanded, the combustion process is disrupted.

Hypersonic flows around a double ellipse are shown in Fig. 6.17, [41]. For the flow with $M_\infty = 25$ and an angle of attack $\alpha = 30°$ Navier–Stokes solutions are computed, where in the left figure a non-equilibrium real gas case ($DAM1 \Longrightarrow O(1)$) and in the right figure a perfect gas case is shown ($DAM1 \Longrightarrow 0$). The influence of the various thermo-chemical assumptions can clearly be seen. The position of the bow shock is much closer to the double ellipse configuration in the non-equilibrium flow case, than in the perfect gas case. This is due the fact that the ratio of the density[11]

[11] ρ the density in front of the bow shock, normally the freestream value and $\hat{\rho}$ the density behind the bow shock.

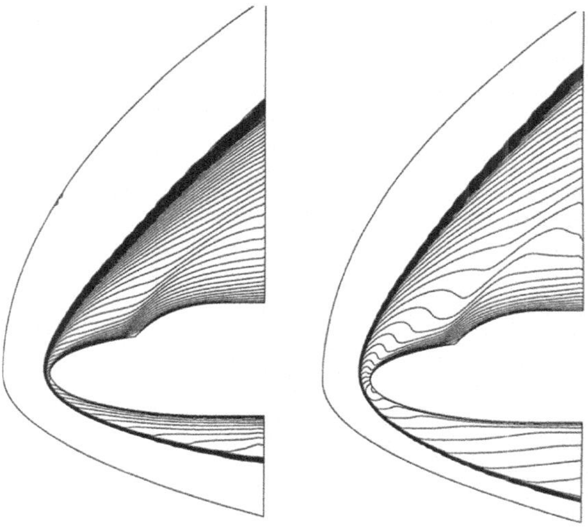

Fig. 6.17 Two dimensional flow past a double ellipse with $M_\infty = 25,\ \alpha = 30°$, Navier–Stokes solution, Iso-Mach lines. Flow with air in chemical non-equilibrium, left; flow with perfect gas assumption (frozen flow), right, [41]

$\hat{\rho}/\rho$ is restricted to the magnitude 6 in the perfect gas case and can be much larger in the non-equilibrium case (depending on the degree of dissociation and chemical reactions). The bow shock is nearest to the configuration for the equilibrium real gas case, ($DAM1 \Longrightarrow \infty$).

In the following we present the example published by Wegener and Buzyna, [42], found in [4].

Given a gas flow around a sphere where the gas consists of the molecule components N_2, N_2O_4, NO_2. The freestream Mach number is $M_\infty = 1.37$. In the regime between the bow shock and the sphere contour the following chemical reaction takes place[12]

$$N_2 + N_2O_4 \lesseqgtr N_2 + 2NO_2$$

The dimensionless shock standoff distance d/R can be described by the four variables

$$\frac{d}{R} = f(M_\infty, \omega_R, \alpha_\infty, DAM1)$$

with ω_R being the mass fraction of the reactants, α_∞ the degree of dissociation. The experience has shown that d/R is in a considerable regime of R only a function of the Damköhler number $DAM1$ ($d/R \approx f(DAM1)$). With the assumptions

[12]When the fluid elements pass the bow shock, the chemical reactions are initiated due to the strong jump of the thermodynamic variables pressure, density and temperature inside the shock.

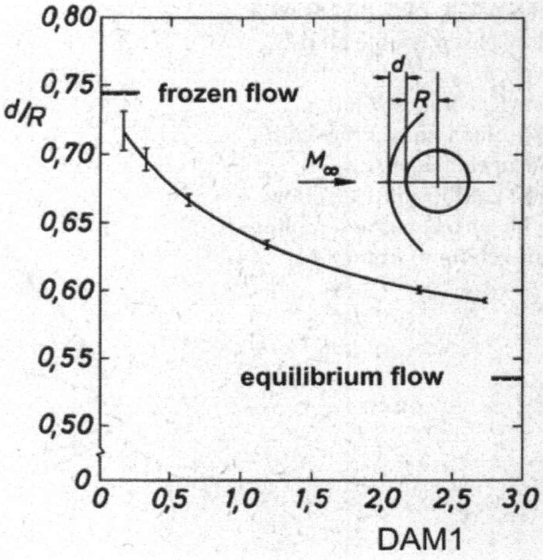

Fig. 6.18 Bow shock distance d/R as function of the first Damköhler number $DAM1$, [42]. Diagram taken from [4]

$M_\infty = 1.37, \omega_R = 0.216, \alpha_\infty = 0.140$ the data plotted in Fig. 6.18 are achieved. Qualitatively the results are the same as the ones presented in Fig. 6.17 in particular d/R is lowest, when the gas tends to be in a thermodynamic equilibrium state.

6.9 The Nusselt Number

Ernst Kraft Wilhelm Nusselt was a German scientist. He was born in Nuremberg on Nov. 25, 1882 and died in Munich on Sept. 1, 1957. He studied mechanical engineering at the Munich Technical University, where he graduated in 1904 and where he prepared his doctoral thesis "Conductivity of Insulating Materials" in 1907. After some temporary positions in Switzerland and Germany he became in 1925 a professor of theoretical mechanics in Munich, where he remained until his retirement in 1952, [43]. The theory of the dimensionless analysis of heat transfer was developed by him. In 1915 he published the famous paper "Das Grundgesetz des Wärmeübergangs" (The Fundamental Law of Heat Transfer), [44]. In this paper he proposed the dimensionless groups of principal parameters in the similarity theory of heat transfer. One of the dimensionless numbers of the transfers of heat bears his name ($Nu \Longrightarrow$ Nusselt number).

Ernst Kraft Wilhelm Nusselt

The Nusselt number Nu describes the convective heat transfer between a surface of rigid material and a fluid, which flows over this surface. It indicates how much the heat transfer to the wall is larger than the conduction in the fluid, Eq. (2.39).

$$Nu = \frac{\alpha L}{\lambda} , \tag{6.13}$$

with α being the heat transfer coefficient, λ the thermal conductivity and L a characteristic length.

Corresponding to Eq. (2.50) the Nusselt number is a function of the seven dimensionless quantities $Ec, Gr, Fr, Pr, Re, \gamma, L/B$. Experience has shown that in a wide range of applications the main dependencies are the Reynolds number Re, the Prandtl number Pr and also the Grashof number Gr, if gravitational forces come into play, often in form of the Rayleigh number $Ra = Gr \cdot Pr$. Therefore the associated Nusselt numbers are mostly constructed as power functions of $Re, Pr, Gr, (Ra = Gr \cdot Pr)$:

$$Nu = Nu\,(Re^l, Pr^m, Gr^n) .$$

The heat flux to a wall is defined by

$$q_w = -\lambda \left(\frac{\partial T}{\partial y}\right)_w = \alpha(T_w - T_{ref}) . \tag{6.14}$$

The first part of Eq. (6.14) is Fourier's law and the second one is Newton's observation (Newton's law of cooling), that the heat transfer depends linearly on an appropriate temperature difference ΔT.[13] If the gradient of the temperature $(\partial T/\partial y)_w$ at the wall is known, the problem is solved. Today we have available modern numerical methods for solving the Navier–Stokes equations, which provide us with this gradient. In earlier days, when it was not possible to determine the term $(\partial T/\partial y)_w$, the empirical relation of Eq. (6.14) was defined in order to have a tool on the basis of the analysis of similarity for the determination of the heat transport of such flow problems.

During the design and development phase of heat exchangers and other technical engineering objects, the method of similarity analysis based on the Nusselt number together with the other similarity parameters found in Eq. (2.50) are still applied today. There are two categories of the heat transfer:
◊ The first one,[14] called forced convection, deals with fluids, which for example flow over plates or pipes or bundle of pipes, partially with different cross-sections, further with flows through pipes, etc. The characteristic point is that there exists

[13]The temperatures forming the difference ΔT are defined with respect to the flow case (heat transfer case) considered, e.g. for T_{ref} often the recovery temperature T_r is applied, which seems to be problematic, in particular when the wall temperature tends towards the recovery temperature.

[14]The second category can be found on Sect. 6.9.1.

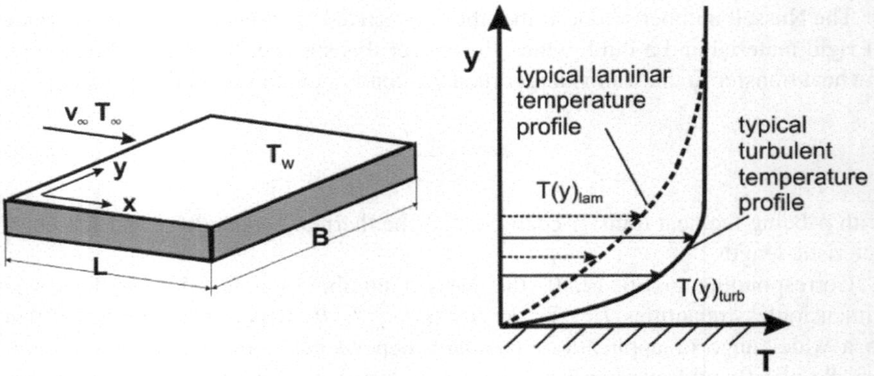

Fig. 6.19 Flow over a heated flat plate (left), Sketch of laminar and turbulent temperature profiles at a surface with heated fluids (right)

a temperature difference between the fluid and the various technical devices. The related heat transfer can be described or at least estimated by the Nusselt number, which for such cases is essentially a function of the Reynolds number and the Prandtl number. Equation (6.14) shows that the heat transfer depends on the temperature gradient normal to the surface. Further this gradient is controlled by the state of the boundary layer either laminar or turbulent, Fig. 6.19 (right). There are a lot of semi-empirical relations of the Nusselt number in particular for

- various flow devices,
- ranges of Reynolds number for laminar, transitional and turbulent flows,
- ranges of Prandtl number for various fluids.[15]

6.9.1 The Nusselt Number for Forced and Free Convection Flows

In the following we exemplarily list some examples of Nusselt numbers. A comprehensive list of Nusselt number approximations[16] can be found in [45].

Flow over a Heated Flat Plate, Fig. 6.19 *(left)*

---

[15]Some Prandtl numbers:
$Pr_{liquid\ metals} << 1$, $Pr_{gases} \approx 0.7$, $Pr_{fluids} \approx 7$, $Pr_{tough\ fluids,\ oils} \approx 70$.

[16]We distinguish between an averaged Nusselt number, called global Nusselt number $Nu_{glob} \equiv Nu_L$, and a local Nusselt number $Nu_{loc} \equiv Nu_x$, which depends on a x-coordinate representing the main stream direction.

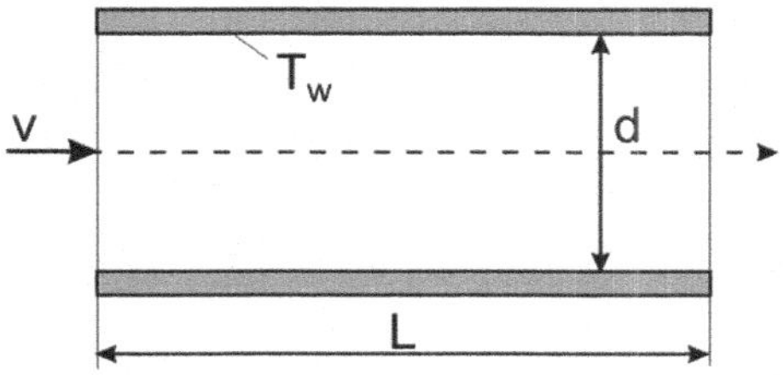

Fig. 6.20 Pipe flow with constant temperature T_w at the inner wall

$$Nu_{glob,lam} = 0.664\ Pr^{1/3}\ Re^{1/2}\ , \quad \text{laminar}$$
$$0.6 < Pr < 2000, \quad Re < 10^5\ , \quad \text{Pohlhausen}$$
$$Nu_{glob,turb} = \frac{0.037\ Pr\ Re^{0.8}}{1 + 2.443\ Re^{-0.1}(Pr^{2/3} - 1)}\ , \quad \text{turbulent}$$
$$0.6 < Pr < 2000, \quad 5 \cdot 10^5 < Re < 10^7 \quad \text{Petukhov/Popov} \tag{6.15}$$

A local consideration of the heat transfer problem can be treated in the same way. Equation (2.50) then gets the form

$$Nu_x = g_2\left(Ec, Gr_x, Fr_x, Pr, Re_x, \gamma, \frac{x}{L}\right)\ , \tag{6.16}$$

and for the laminar case a possible semi-empirical formula reads

$$Nu_{x,lam} = 0.332\ Pr^{1/3}\ Re_x^{1/2}\ , \quad \text{laminar}$$
$$0.6 < Pr < 2000, \quad Re_x < 5 \cdot 10^5\ . \quad \text{Pohlhausen} \tag{6.17}$$

Pipe Flow, Fig. 6.20

$$Nu_{glob,lam} = \left(Nu_1^3 + 0.7^3 + (Nu_2 - 0.7)^3 + Nu_3^3\right)^{1/3}\ , \quad \text{laminar}$$
$$0 < Re\ Pr\frac{d}{L} < \infty\ , \quad \text{Martin}$$
$$Nu_1 = 3.66,\ Nu_2 = 1,615\left(Re\ Pr\ \frac{d}{L}\right)^{1/3}\ ,$$
$$Nu_3 = \left(\frac{2}{1 + 22Pr}\right)^{1/6}\left(Re\ Pr\frac{d}{L}\right)^{1/2}\ ,$$

for the inlet regime, otherwise $Nu_3 = 0$,

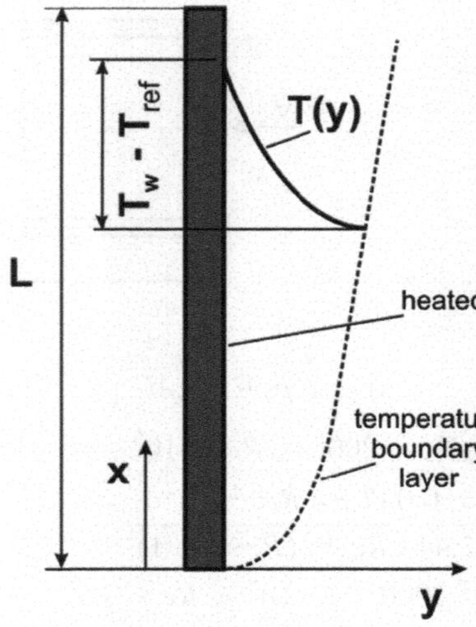

Fig. 6.21 Free convection flow driven by temperature differences between a heated vertical wall and a fluid (see also Fig. 6.4)

$$Nu_{glob,turb} = 0.0214\left(Re^{0.8} - 100\right)Pr^{0.4}\left(1 + \left(\frac{d}{L}\right)^{2/3}\right), \quad \text{turbulent}$$

$$0.5 < Pr < 1.5, \quad 2300 < Re < 10^6, \quad \text{Gnielinski} \tag{6.18}$$

◊ The second category, called free convection, deals with fluids, whose convective motion is driven by gravitational forces (lift forces) due to density differences generated by temperature differences $\Delta T = T_w - T_{ref}$ between a wall T_w and a reference value of the fluid T_{ref}, see also the paragraph about the Grashof number, Sect. 6.4.

Free Convection Flow Along a Vertical Wall, Fig. 6.21

As already mentioned, for free convection flows the Grashof number plays a role. In the following relations the Grashof number appears and sometimes also the Rayleigh number, which is defined by $Ra = Gr \cdot Pr$, see Sect. 6.16.

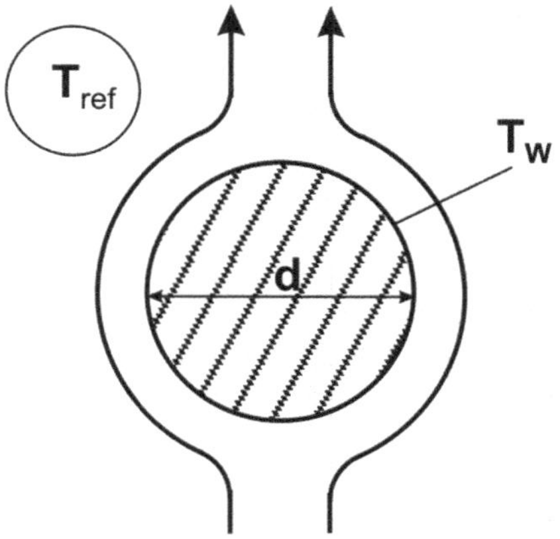

Fig. 6.22 Free convection flow driven by temperature differences between a heated horizontal cylinder and a fluid

$$Nu_{glob} = \left(0.825 + 0.387\,[Ra_L f_1(Pr)]^{\frac{1}{6}}\right)^2 , \qquad \text{laminar/turbulent}$$

$$\text{where} \quad f_1(Pr) = \left(1 + \left(\frac{0.492}{Pr}\right)^{\frac{9}{16}}\right)^{-\frac{16}{9}} ,$$

$$0.001 < Pr < \infty, \quad 0.1 < Ra_L < 10^{12} . \qquad \text{Churchill / Chu} \tag{6.19}$$

A possible relation for the local Nusselt number is given by

$$Nu_x = 0.508\, Pr^{\frac{1}{2}} (0.952 + Pr)^{-\frac{1}{4}}\, Gr_x^{\frac{1}{4}} ,$$
$$0.01 < Pr < 1000, \quad 10^4 < Ra_L < 10^9. \tag{6.20}$$

Free Convection Flow Past a Heated Horizontal Cylinder, Fig. 6.22

$$Nu_{glob} = \left(0.752 + 0.387\,[Ra_L f(Pr)]^{\frac{1}{6}}\right)^2 , \qquad \text{Churchill / Chu}$$

$$\text{where} \quad f(Pr) = \left(1 + \left(\frac{0.559}{Pr}\right)^{\frac{9}{16}}\right)^{-\frac{16}{9}} , \quad L = \frac{\pi}{2} d ,$$

$$0 < Pr < \infty . \tag{6.21}$$

6.10 The Stanton Number

Sir Thomas Edward Stanton was a British engineer, born at Atherstone, England, on Dec. 12, 1865. He worked during his study (completed in 1891) and some years afterwards at the laboratory of Osborne Reynolds in Manchester. In 1899 he became a post as professor of engineering at Bristol University and remained there until his retirement in 1930, one year before his death. His scientific work encompassed activities in fluid dynamics, in particular the issues of flows with friction, and the problems of heat transfer of fluids to rigid walls (flows around configurations of flight vehicles, pipe flows, flows through devices of process engineering, etc.). The dimensionless number concerning the ratio of convective transferable heat to total transferable heat bears his name the Stanton number St.

Sir Thomas Edward Stanton

The Stanton number is defined by

$$St = \frac{\alpha}{\rho_\infty\, u_\infty\, c_p}\,, \tag{6.22}$$

with α the heat transfer coefficient, ρ_∞ the density, u_∞ the velocity of the freestream and c_p the specific heat.[17] The Stanton number names the ratio between the convective transported heat of the flow process to the total transported heat. Another interpretation of the Stanton number states, that it is the ratio of the process dynamics compared to the capability to store heat energy, which can be formulated as: The greater the Stanton number the faster merges the heat to the wall (or the body). The Stanton number can be constituted as function of the Nusselt number (Nu), the Prandtl number (Pr), the Reynolds number (Re) or alternatively the Péclet Pe number[18] and one obtains

$$\frac{\alpha\, L}{\lambda_\infty} \cdot \frac{\lambda_\infty}{\mu_\infty\, c_p} \cdot \frac{\mu_\infty}{\rho_\infty\, u_\infty\, L} = Nu \cdot \frac{1}{Pr} \cdot \frac{1}{Re} = \frac{Nu}{Pe} = St\,. \tag{6.23}$$

In fluid dynamics the Stanton number with Eq. (6.14) has the form

$$St = \frac{\alpha}{\rho_\infty\, u_\infty\, c_p} = \frac{\alpha(T_w - T_r)}{\rho_\infty\, u_\infty\, c_p(T_w - -T_r)} = \frac{q_w}{\rho_\infty\, u_\infty\, (h_w - h_r)}\,. \tag{6.24}$$

[17]The specific heat is denoted by c_p for gases at constant pressure and c for fluids.

[18]See Eqs. (2.12), (2.39), (2.46), (3.9).

6.10.1 The Relevance of the Stanton Number for Hypersonic Flows

During the time period 1986–1993 the European Space Agency ESA had initiated the Hermes programme for the development of a manned space transportation system, essentially in the form of a winged re-entry vehicle, in order to allow Europe an autonomous access to space. One of the main challenges of the aerothermodynamic discipline (hypersonic flows) is the prediction of the heat loads during the re-entry phase. Numerous activities on the experimental as well as the numerical side were undertaken in order to understand the physics of the various kinds of heat transfer in hypersonic flows. In the frame of an accompanying European research program numerous numerical methods for solving the three dimensional Navier–Stokes equations for flows around realistic configurations were developed and their capabilities were demonstrated by presentations in some workshops.[19] One of these test configurations was the Blunt Delta Wing[20] with a shape of the windward side close to that of the Hermes configuration.

The absolute values of the Stanton number in the graphs of Fig. 6.23 show a remarkable increase[21] of St along the leading edge, which indicates an increase of the heat transfer rate. That is what one physically expects, because the thermal boundary layer thickness decreases due to the flow expansion around the leading edge.

A second workshop test case was the hypersonic flow over a two dimensional ramp. The physical reason for this test case was to understand the mechanisms of heat transfer in corners (with recirculation regimes) and behind embedded shock waves in hypersonic flows.[22] Various experimental and numerical investigations were performed, [46]. We show here the results of the experimental test campaign, Fig. 6.24, [48]. The Stanton number distributions for the two different wall temperatures exhibits the same qualitative behavior, but have different levels, Fig. 6.24 (below).

Generally the usefulness of the Stanton number as defined by Eq. (6.24) must be seen critical for hypersonic flight. On the one hand in experiments the Stanton number is only a measure for cold wall heat transfer rates, since the model is usually not heated up in wind tunnel tests to temperatures, which are realistic in hypersonic flight (normally greater than 1000 K). On the other hand the surface radiation cooling, which reduces the wall temperatures dramatically, is not taken into account in the above described flow cases, since they had cold walls, see [24, 39].

---

[19] These were the famous Antibes Workshops, [41, 46].

[20] This configuration was designed by the Space Department of Messerschmitt—Bölkow—Blohm, Ottobrunn, Germany.

[21] The negative values of the Stanton number in Fig. 6.23 (below) are due to another definition of St compared to Eqs. (6.22)–(6.24), namely $St = \frac{-\lambda \frac{\partial T}{\partial n}}{\rho_\infty u_\infty (h_0 - h_w)}$, which generates a negative sign.

[22] These are some of the most critical aspects of hypersonic flight as the Columbia disaster has shown in 2003.

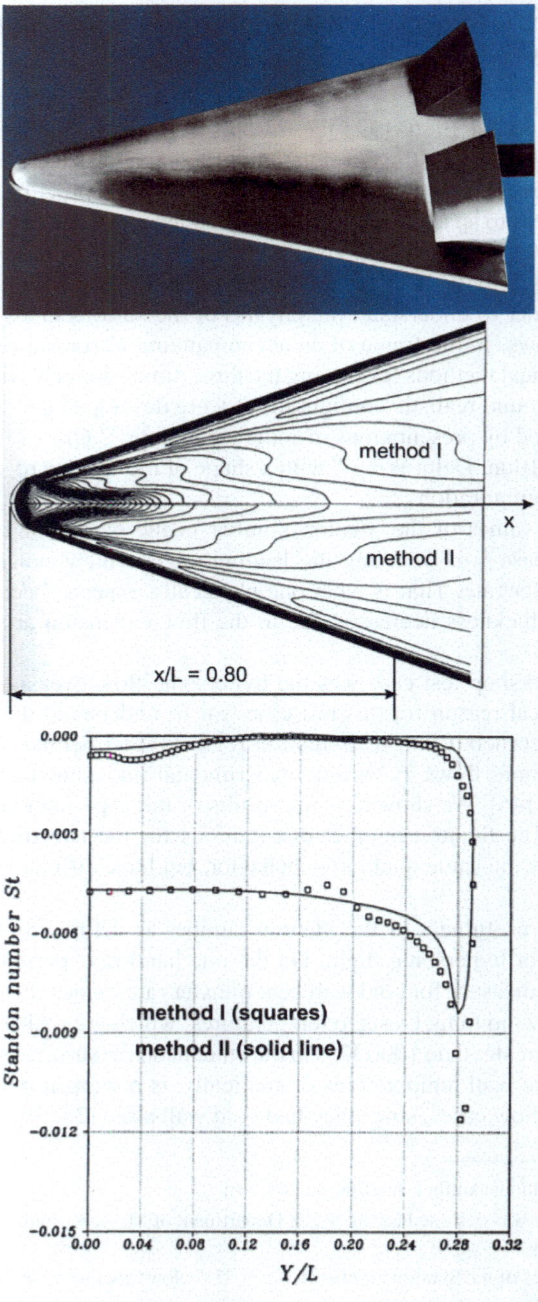

Fig. 6.23 Hypersonic flow past a Blunt Delta Wing. Results of two different Navier–Stokes solutions [47]. $M_\infty = 8.7$, $Re_\infty/m = 2.25 \cdot 10^6$, $\alpha = 30°$, $T_{wall} = 300$ K, $T_\infty = 55$ K. Wind tunnel model including deflected flaps (above). Stanton number contours of the leeward side (middle). Stanton number at the crossflow section x/L = 0.8 (below)

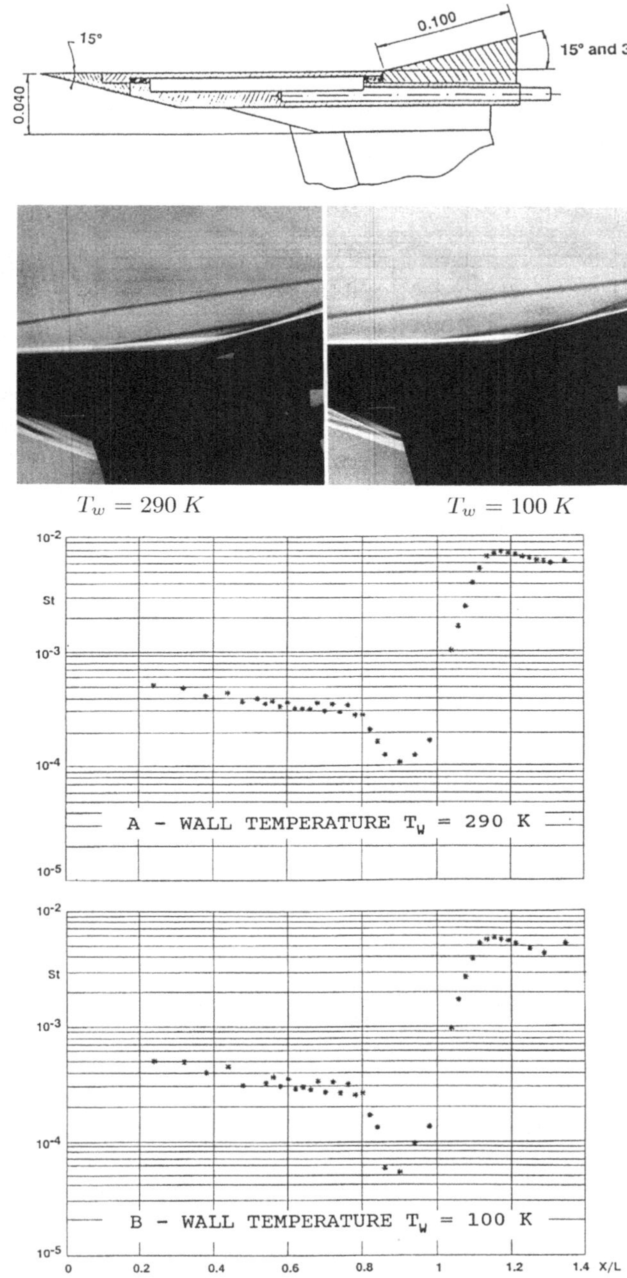

Fig. 6.24 Hypersonic flow over a 2D ramp with a 15 degree ramp angle. Experimental investigations for two different wall temperatures T_w, [48]. $M_\infty = 10$, $Re_L = 2.3 \cdot 10^6$. Flow model (above), Schlieren photographs (middle), surface Stanton number distribution for $T_w = 290$ and 100 K (below)

6.11 The Lewis Number

Warren Kendall Lewis (August 21, 1882–March 9, 1975) was an US-American scientist. He studied chemical engineering at the Massachusetts Institute of Technology (MIT) in Boston, USA. In the frame of a post graduate study of physical chemistry in Breslau, Germany, he received his Sc.D. degree. Back to MIT he became in 1914 a full professor and was the first head (1920) of the newly established Department of Chemical Engineering. He remained at MIT until his retirement in 1948 and worked further as professor emeritus at MIT until his death in 1975, [49]. His scientific work was mainly related to chemical engineering, where he published the seminal paper on the Theory of Fractional Distillation, [50]. During World War II he chaired several committees for surveying the progress in the Manhattan project.

Warren Kendall Lewis

The dimensionless number, which relates the heat transport by conduction to the heat transport by mass diffusion bears his name.

The Lewis number is defined by

$$Le = \frac{\lambda}{\rho\, c_p\, D} = \frac{k}{D}\,, \tag{6.25}$$

and is a pure substance based dimensionless number, with k being the thermal diffusivity, D the mass diffusivity, λ the conductivity, ρ the density and c_p the specific heat at constant pressure. The Lewis number can also be expressed as the ratio of the Schmidt number Sc to the Prandtl number Pr, $Le = Sc/Pr$, Sect. 6.12.

6.11.1 A Typical Application Case of the Lewis Number

The Lewis number characterizes gas mixtures. The typical case is the combustion of gas mixtures, where the structure of the flame (flame stabilization, flame propagation, influence of mass diffusion, etc.) is object of the investigation, [51]. Gas mixtures can have Lewis numbers smaller than unity ($Le < 1$), e.g. lean hydrogen/air or hydrogen/chlorine $Le \approx 0.4$, close to unity ($Le \approx 1$), e.g. lean methane/air, and larger than unity ($L > 1$), e.g. benzol/air. It can be expected that mass diffusion plays only a role for $Le < 1$, whereas this effect can be neglected for Lewis numbers towards one and beyond.

In Table 6.3 Lewis numbers for some often used gas mixtures and gaseous species are listed.

Table 6.3 Lewis numbers for some gas mixtures (left), and for gaseous species (right), reference temperature $T_{ref} = 273$ K, reference pressure $p_{ref} = 1$ atm, [52, 53]

Gas mixture	Lewis number	Gas	Lewis number
Ethanol—air	1.84	H_2O	0.80
Benzol—air	2.49	CO_2	1.24
Methanol—air	1.41	H_2	0.26
Water vapor—air	0.81	O_2	0.99
Ethanol—CO_2	1.32	N_2	0.86
Benzol—CO_2	1.72	CH_4	0.88

6.12 The Schmidt Number

Ernst Heinrich Wilhelm Schmidt was a German scientist and engineer, whose field of activities was the thermodynamics and in particular the phenomena of heat and mass transfer. Born in Vögelsen, Germany on Feb. 11, 1892, he studied civil and electrical engineering at the Technical Universities of Dresden and Munich, Germany. He graduated there, after World War I, in 1919 and prepared his doctoral thesis in 1920. He was appointed as a professor at the University of Danzig (1925). In 1937 he became the director of the Institute of Propulsion at the Aeronautical Research Establishment in Braunschweig, and went in 1952 to the University of Munich as the successor on the chair of E.K.W. Nusselt, where he stayed until his retirement in 1961, [54]. On Jan. 22, 1975 he died in Munich. Since E. Schmidt has investigated and clarified a lot of the properties regarding the analogies between momentum, heat and mass diffusion, the dimensionless number describing the ratio between the diffusive momentum transport and the diffusive mass transport carries his name (Sc).

Ernst Heinrich Wilhelm Schmidt

Table 6.4 Schmidt numbers of some fluids added in lean concentration to air and to water, [55]. $p_{ref} = 0.1$ MPa

Added fluids	Into air Sc	Into water Sc
Hydrogen	0.22	154
Oxygen	0.75	361
Carbon dioxide	0.96	461
Methanol	1.00	564
Ethanol	1.30	705

The Schmidt number is a substance based dimensionless number and is defined by

$$Sc = \frac{\nu}{D} = \frac{k}{D} \cdot \frac{\nu}{k} = Le \cdot Pr\;, \tag{6.26}$$

with ν being the kinematic viscosity, D the coefficient of mass diffusion, k the thermal diffusivity.

6.12.1 *The Role of the Schmidt Number for Mass Diffusion in Fluids and Gases*

The importance of the Schmidt number can be illustrated by the following examples. First of all we have a look at Table 6.4, where the Schmidt numbers for some fluids injected on the one hand into air and on the other hand into water are listed.

This list reflects, that fluids, which are added to air, have Schmidt numbers of order unity ($Sc \Rightarrow \mathrm{O}(1)$), while fluids in water have Schmidt numbers of order 1000 ($Sc \Rightarrow O(10^3)$). In the first case this has the consequence that strong mass diffusion occurs as it is demonstrated in Fig. 6.25, where smoke plumes around a car in a wind tunnel are shown. It can clearly be seen, that in the aft part of the car the plumes are widened, indicating the high degree of mass diffusion.

Fig. 6.25 Automobile in a wind tunnel, [56]. Smoke lines in the symmetry plane exhibit the flow behavior. Schmidt number Sc amounts approximately to 1

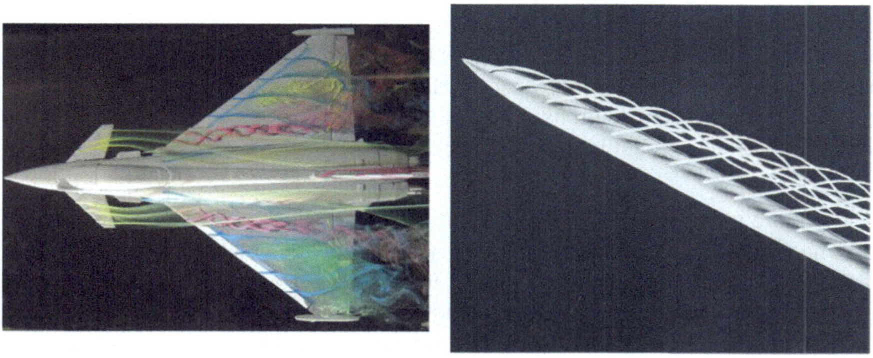

Fig. 6.26 Combat aircraft in a water tunnel, [57]. Color filaments exhibiting the flow structure, in particular the vortex formations and their break down, (left). Vortex pair behind an inclined (30°) long ogive-cylinder in a water tunnel. $Re = 400$ based on the cylinder diameter, [27], (right). The Schmidt number Sc amounts in both cases to approximately 10^3

Alternatively color filaments in water have a high Schmidt number—the mass diffusion is low—with the consequence that the filaments are longer stable and expand much later,[23] Fig. 6.26.

[23] The fraying of the color filaments on the rear part of the upper side (leeward side) of the combat aircraft is due to the vortex break down and the subsequent turbulence onset.

6.13 The Prandtl Number

Ludwig Prandtl was born on Feb. 4, 1875 in Freising and died on Aug. 15, 1953 in Göttingen. He studied engineering sciences and prepared his doctoral thesis (Dr.phil.) in 1900 in Munich. At the age of 26 years he was appointed as a professor for mechanics at the university of Hannover (1901). Afterwards the famous mathematician Felix Klein had persuaded Ludwig Prandtl to accept a post as professor in Göttingen (1904), where he remained until his death in 1953. Already in 1904 Ludwig Prandtl had published the seminal article "Über Flüssigkeitsbewegung bei sehr kleiner Reibung" (About fluid motion with very small friction.), [58], where the concept of the boundary layer (Grenzschicht) was introduced. He provided further outstanding contributions in the fields of gasdynamics, turbulence modelling, wing theory, similarity theory, etc., [59].

Ludwig Prandtl

Theodore von Kármán stated about him (1954), [60]:

"The man who gave modern wing theory its practical mathematical form was one of the most prominent representatives of the science of mechanics, and especially fluid mechanics, of all time, Ludwig Prandtl".

The dimensionless number describing the ratio of the diffusive momentum transport and the convective heat transport is termed by the Prandtl number Pr.

The Prandtl number, already defined by Eqs. (2.46) and (4.20), reads

$$Pr = \frac{\mu\, c_p}{\lambda} = \frac{\nu}{k}\,, \tag{6.27}$$

with $\mu = \rho\,\nu$ being the dynamic viscosity, ν the kinematic viscosity, λ the heat conduction coefficient, $k = \lambda/\rho c_p$ the thermal diffusivity, ρ the density and c_p the specific heat at constant pressure. Since the kinematic viscosity ν stands for the momentum transport and the thermal diffusivity k for the heat (energy) transport, which represent the velocity and the temperature fields, the Prandtl number is a link between both.

6.13.1 Some Considerations Regarding the Significance of the Prandtl Number

Prandtl numbers depend, with respect to the material considered, on the temperature and the pressure (in case of a gas). To demonstrate that we have a look on Fig. 6.27, where the Prandtl number for air as function of the temperature T and the pressure p

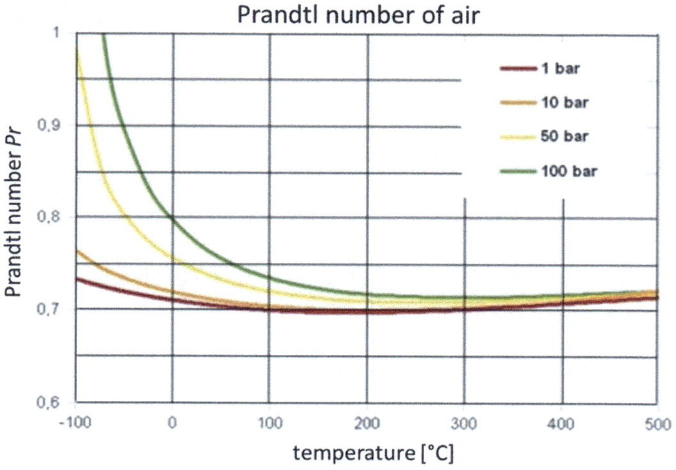

Fig. 6.27 Prandtl number for air as function of temperature and pressure

is plotted. Generally for gases the influence of the temperature on the Prandtl number increases with lower values of T and higher values of p.

In Table 6.5 four different groups of materials are listed. Regimes of Prandtl number values including some examples are defined and show a large value band. The Prandtl number is also a measure of the ratio of the boundary layer thicknesses, either δ_m for the momentum diffusivity or δ_T for the thermal diffusivity. For arbitrary gas flows with $Pr = 1$, δ_m and δ_T are of the same order of magnitude.[24] Small Prandtl numbers $Pr \ll 1$ (liquid metals) lead to $\delta_T \gg \delta_m$, which means that the viscous boundary layer can be neglected in further considerations. On the other hand $Pr \gg 1$ (oils) implies $\delta_T \ll \delta_m$, Fig. 6.28.

Considering the heat transfer for the flow along a flat plate, the relation $\delta_T \approx \delta_m Pr^{-1/3}$ is valid for laminar flow. In case of water with $Pr = 7$ this means $\delta_m \approx 1,912\,\delta_T$, [62, 63].

There exists also a definition for a turbulent Prandtl number. The apparent turbulent shear stress is defined by $-\overline{u'\,v'} = \nu_t \dfrac{\partial \overline{u}}{\partial y}$ and the apparent turbulent heat flux by $-\overline{v'\,T'} = k_t \dfrac{\partial \overline{T}}{\partial y}$, where ν_t is called the momentum eddy viscosity and k_t the heat transfer eddy viscosity. The total momentum diffusivity is $\nu_{total} = \nu + \nu_t$ and the total thermal diffusivity amounts to $k_{total} = k + k_t$. The turbulent Prandtl number then reads $Pr_t = \dfrac{\nu_t}{k_t}$, which is not a dimensionless number in the sense of the dimensional analysis, since it is built with the quantities of the flow. When the Reynolds analogy is assumed it is found that the velocity field and the temperature field are similar, which would mean $Pr_t \approx 1$. But the reality has revealed that there

[24]For the flow along a flat plate with $T_w = const$, δ_m and δ_T are identical for $Pr = 1$.

Table 6.5 Prandtl number regimes for various materials and some samples, [7, 18, 61], (The temperature is $T = 20\,°C$, if not other specified)

Materials	Pr regime	Examples
Liquid metals	$0.005 < Pr < 0.04$	Liquid sodium Na (0.005)
Gases ($p = 0.1$ MPa)	$0.7 < Pr < 1.5$	Water vapor (1.12 at $T = 100\,°C$), Air (0.7), CO_2 (1.38 at $T = 0\,°C$)
Fluids	$2 < Pr < 80$	Water H_2O (7.0), benzol C_6H_6 (7.5), Ethanol C_2H_5OH (18.8)
Oils	$80 < Pr < 12000$	Ricinus oil (10500), glycerine (11340), Transformer oil (480)

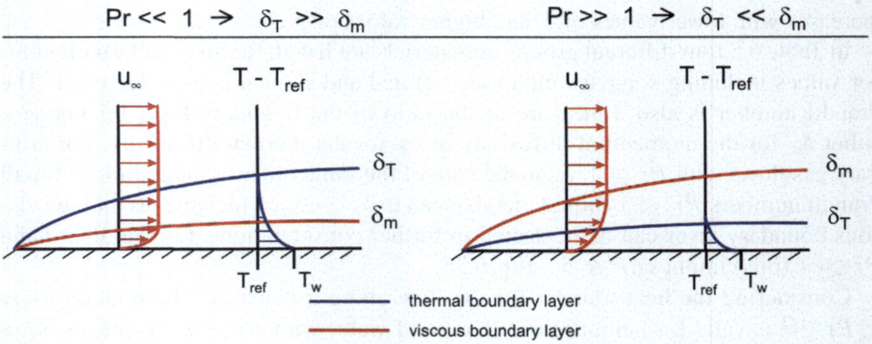

Fig. 6.28 Laminar boundary layers: relations of boundary layer thicknesses, either viscous or thermal, for extreme Prandtl numbers

is no similarity of the velocity and temperature fields and that therefore $Pr_t \approx 1$ is not valid. Experimental data yield for airflows a turbulent Prandtl number range of $0.7 < Pr_t < 0.9$. Further, there exist a lot of theoretical and/or semi-empirical proposals for the determination of the turbulent Prandtl number, see [64].

6.14 The Fourier Number

Jean-Baptiste-Joseph Fourier

Jean-Baptiste-Joseph Fourier was born on March 21, 1768 in Auxerre, France and died on May 16, 1830 in Paris. In 1795 he got a position as teacher at the École Polytechnique in Paris. He was interested in the French Revolution with the consequence that he was imprisoned for a short time. In 1798 he went to Egypt, due to an offer of Napoleon, as a military scientific adviser. Afterwards, when he returned to Paris, Napoleon has committed him several administrative positions, among others as Prefect of the *Isère Départment* with headquater in Grenoble (1801), [65, 66]. In 1807 he presented in Paris the article "On the Propagation of Heat in Solid Bodies". After the defeat of Napoleon in 1815 he resumed his research on heat and published the book "Analytical Theory of Heat" (1822), [67].

He proposed to solve his differential equation for the propagation of heat (Fourier's law) with a series of sines and cosines (Fourier series, later Fourier transform), [68].

The dimensionless number describing the ratio of rate of heat conduction to the rate of heat storage bears his name.

The Fourier number given by Eq. (3.9) reads

$$Fo = \frac{\lambda\, t}{\rho\, c_p\, l^2} = \frac{k\, t}{l^2}\,, \tag{6.28}$$

with λ being the heat conduction coefficient, ρ the density, c_p the specific heat at constant pressure, k the thermal diffusivity, l a reference length of a physical problem considered and t the propagation time. The Fourier number describes the process of unsteady heat conduction in all matters. It has the form of a dimensionless time $Fo = \frac{k\, t}{l^2} = t^*$. Further it can also be interpreted as the ratio of current time to time to reach the steady state.

Besides the standard Fourier number also a Fourier number is defined for the mass diffusion and one has

$$\bar{Fo} = \frac{D\, t}{l^2}\,, \tag{6.29}$$

where the thermal diffusivity k of Eq. (6.28) is replaced by the mass diffusivity D.

6.14.1 The Process of Heat Conduction Described by the Fourier Number

The general energy equation is given by Eqs. (3.4), (4.10) and reformulated by Eq. (3.8). When we drop in this equation the convective terms, the power of compression due to the pressure term, the dissipation term and the heat source terms, we obtain the equation for unsteady heat conduction.

$$\rho\, c_p \frac{\partial T}{\partial t} = \frac{\partial}{\partial x}\left(\lambda \frac{\partial T}{\partial x}\right) + \frac{\partial}{\partial y}\left(\lambda \frac{\partial T}{\partial y}\right) + \frac{\partial}{\partial z}\left(\lambda \frac{\partial T}{\partial z}\right) = div\,(\lambda\, grad\, T)\,. \tag{6.30}$$

As an example we consider the 1-D heat conduction in a flat plate of infinite width with the thickness $2s$, which is surrounded by a fluid in a large basin of constant temperature T_∞. The flat plate has the initial temperature T_i, Fig. 6.29. For this 1-D example with $\lambda \neq \lambda(x)$ we find

$$\begin{aligned} \rho\, c_p \frac{\partial T}{\partial t} &= \lambda \frac{\partial^2 T}{\partial x^2}\,, \\ \frac{\partial T}{\partial t} &= \frac{\lambda}{\rho\, c_p}\frac{\partial^2 T}{\partial x^2} = k \frac{\partial^2 T}{\partial x^2}\,, \\ k &= \frac{\lambda}{\rho\, c_p} \qquad \text{the thermal diffusivity}\,. \end{aligned} \tag{6.31}$$

The general solution of Eq. (6.31) is given by Fourier series, [69],

$$T(x,t) = \sum_{n=1}^{\infty} (C_1 cos(Bx) + C_2 sin(Bx))\, e^{-kC_3 t} \tag{6.32}$$

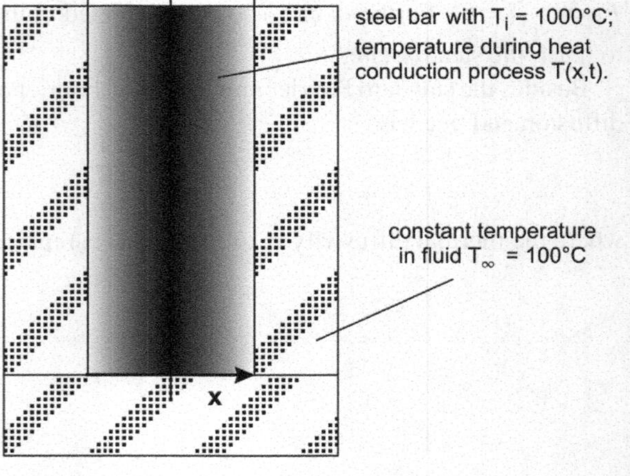

Fig. 6.29 Sketch of a flat plate (steel bar) of infinite width and a thickness of $2s$ with an initial temperature $T_i = 1000\,°C$ plunged in a fluid with the constant temperature $T_\infty = 100\,°C$

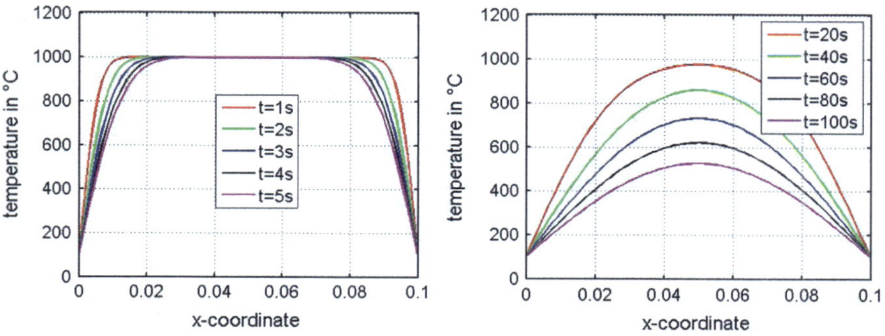

Fig. 6.30 Unsteady, 1-D heat conduction in a flat plate. The results are obtained by solving numerically Eq. (6.33). The heat transport (conduction) to the wall has not reached the middle of the bar ($x = s$) during the first five seconds (left diagram). For times larger than 20 seconds ($t > 20s$) the heat transport has captured the whole thickness of the bar (right diagram)

and with the initial and boundary conditions,[25] which specify the constants B, C_1, C_2, C_3 of our example, we obtain, [70],

$$\frac{T(x,t) - T_\infty}{T_i - T_\infty} = \sum_{n=1}^{\infty} C_2 sin\left(\frac{n\pi}{2s}x\right) e^{-(\frac{n\pi}{2})^2 Fo}\,, \tag{6.33}$$

with $Fo = kt/s^2$ being the Fourier number and $C_2 = \dfrac{2}{n\pi}(1-(-1)^n)$.

The thickness of the flat plate was chosen to be $2s = 10$ cm and the thermal diffusivity of steel is $k \approx 10^{-5}\ \mathrm{m^2/s}$. We have evaluated Eq. (6.33) for $0 < x < 2s$ and in a first campaign for $t = 1s, 2s, 3s, 4s, 5s$, Fig. 6.30 (left), and in a second campaign for $t = 20s, 40s, 60s, 80s, 100s$, Fig. 6.30 (right). The Fourier numbers were $Fo = 0.004, 0.008, 0.012, 0.016, 0.020$ and $Fo = 0.08, 0.16, 024, 0.32, 0.40$ respectively.

For low Fourier numbers Fo and small times t the penetration depth of the heat conduction has not reached the midpoint $x = s$ of the bar, which means that the Fourier series converge only very slowly. This changes for larger Fourier numbers, where sometimes just one term is sufficient.

Remark For low Fourier numbers another solution of Eq. (6.31) can be applied, [69, 71], when the penetration depth is $x_{pen} \leqq s$, namely

$$T(x,t) = T_\infty + (T_i - T_\infty)\, erf\left(\frac{x}{2\sqrt{kt}}\right)\,, \tag{6.34}$$

with $erf(x/2\sqrt{kt})$ being the Gauss' error function and

[25] The heat transfer coefficient between fluid and plate goes to infinity, from which follows, that the temperature of the fluid is equal to the temperature of the wall!.

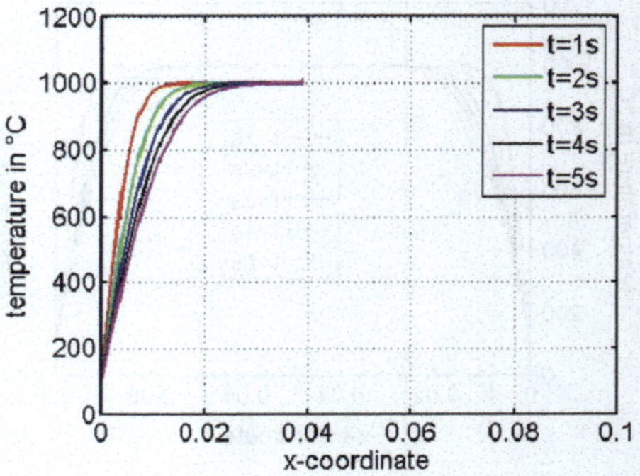

Fig. 6.31 Unsteady, 1-D heat conduction in a flat plate. The first five seconds of the heat transport process, computed with Eq. (6.34), are indicated

$$erf\left(\frac{x}{2\sqrt{kt}}\right) = \frac{2}{\sqrt{\pi}}\sum_{n=0}^{\infty}\frac{(-1)^n\left(\frac{x}{2\sqrt{kt}}\right)^{2n+1}}{n!\,(2n+1)}\,. \tag{6.35}$$

We have evaluated this function for our problem above. The result is presented in Fig. 6.31 and is nearly identical to Fig. 6.30 (left).

It may be of interest to know after what time the propagation depth has reached the centerline $x = s$. We did this investigation by a numerical iteration process and have found that $x = s$ was catched for a time $8s \leq t \leq 9s$. The iteration process was stopped for $T \geq 999.9\,°C$.

6.15 The Strouhal Number

Vincent Strouhal, a Czech scientist, was born in Seč on April 10, 1850 and died in Prague on January 26, 1922. He was an experimental physicist specialized on hydrodynamics. In 1882 he was appointed as full professor for experimental physics at the newly established faculty of physics,—at which he was one of the founder—, at the University of Prague, Czech Republic. In 1878 he presented at the "Physikalisch-Medizinischen Gesellschaft, Würzburg" the paper "Über eine besondere Art der Tonerregung" (About a specific type of sound excitation.), [72], where for the first time the relation was applied, which has afterwards born his name, the Strouhal number.

He wrote several textbooks in the context of the physical disciplines mechanics, acoustics, thermodynamics and optics, [73].

Vincent Strouhal

The dimensionless number given by Eqs. (2.10), (3.2), (4.15) is the Strouhal number

$$Str = \frac{l}{ut} = \frac{fl}{u}\,, \tag{6.36}$$

with l being a reference length, u the flow velocity, t a characteristic time (periodic time) or $f = 1/t$ a characteristic frequency (periodic frequency) of the physical phenomenon considered.

Note The Strouhal number can be obtained also from the continuity equation using the same process as in Sect. 4.3. For the unsteady one dimensional case we find

$$\frac{\partial \rho}{\partial t} + \rho \frac{\partial u}{\partial x} + u \frac{\partial \rho}{\partial x} = 0\,, \tag{6.37}$$

and with $u' = \frac{u}{\bar{v}}$, $\rho' = \frac{\rho}{\bar{\rho}}$, $t' = \frac{t}{\bar{t}}$, $x' = \frac{x}{l}$ we obtain

$$\underbrace{\frac{l}{\bar{v}\bar{t}}}_{Str} \frac{\partial \rho'}{\partial t'} + \rho' \frac{\partial u'}{\partial x'} + u' \frac{\partial \rho'}{\partial x'} = 0\,. \tag{6.38}$$

6.15.1 Three Interesting Examples Regarding the Application of the Strouhal Number

(1) Flows past obstacles produce various forms of separation structures, which depend on the Reynolds number of these flows. Only in the case of $Re \Rightarrow 0$, creeping flow, no separation will occur. For example, for the two dimensional flow past a circular cylinder the well-known v. Kàrmàn vortex street in its pure formation arises for Reynolds numbers $50 \lessapprox Re \lessapprox 300$, Fig. 6.32. The quantity of the Strouhal number Str starts at $Re \approx 50$ with $Str \approx 0.14$ and becomes constant with increasing Reynolds number, $Str \approx 0.21$ for $10^3 \lessapprox Re \lessapprox 10^5$. Flows around other shapes than circular cylinders, e.g. cylinders with cross-sections like a wedge, a cuboid or a projectile, have Strouhal numbers up to $Str \approx 0.27$, [74, 75].

(2) We consider a wire with a diameter of 1 cm, which is situated in a wind flow with the velocity $u = 22\,\text{m/s}$. Assuming that the Strouhal number amounts to $Str = 0.2$, we obtain for the separation frequency, which is equivalent with the excitation frequency

$$f = Str\,\frac{u}{d} = 0.2\,\frac{22}{0.01} = 440\,\text{Hz}\,,$$

and the wire is singing.

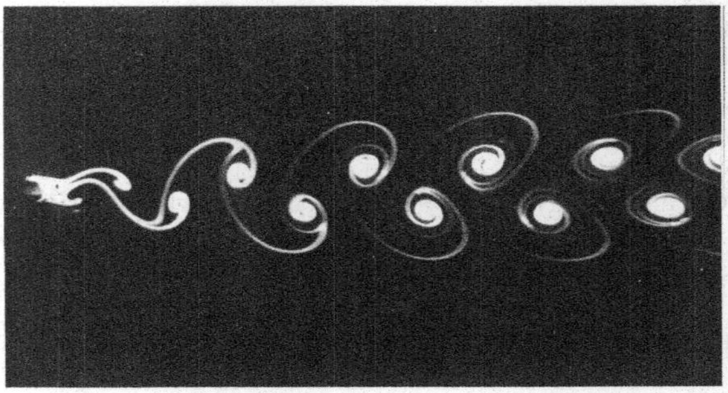

Fig. 6.32 Two dimensional flow around a circular cylinder in water with $Re = 105$ and a Strouhal number of $Str \approx 0.19$. Generation of a v. Kàrmàn vortex street, photograph by Sadatoshi Taneda, [27]

(3) On Nov. 7, 1940 a slight storm was blowing with continuous wind speeds of $67.57\ \mathrm{km/h} = 18.77\ \mathrm{m/s}$ in Tacoma, Pierce County, Washington, USA. And it happened that at the roadway of the Tacoma Narrows Bridge periodic flow separation occurred, which led to an amplification of torsional oscillations of the roadway, which were not controllable. The roadway was lifted up to about 9 m, Fig. 6.33 (above), and twisted by angles of $\pm\,45°$, Fig. 6.33 (below). After 45 min a roadway piece of 200 m length fell down and later the whole bridge collapsed, Fig. 6.34.

The Tacoma Narrows Bridge desaster was investigated very intensively. As mentioned, at the time of the failure the wind speed has amounted to $u = 67.57\ \mathrm{km/h} = 18.77\ \mathrm{m/s}$ (42 mph). The relevant reference length of the roadway was $l = 2,438$ m (8 ft).[26] The whole experimental experience of this type of flow indicates a Strouhal number of $Str \approx 0.2$. The observation of a person (F. B. Farquharson of the University of Washington) at the time of the break down of the bridge was, that the roadway did oscillate with a frequency of approximately 0.2 Hz. The frequency evaluated by the Strouhal number relation, which refers to the corresponding v. Kàrmàn vortex street, was $f = Str\ u/l = 0.2 \cdot\ 18.77/2.438 = 1.54$ Hz. This is far away from the observed value (0.2 Hz).

Sometimes the disaster was physically explained by a resonance effect between the frequency of the v. Kàrmàn vortex street and the natural frequency of the structure. But due to the facts described above, strong questions arise that this could be correct.

Some investigations of engineers and physicists come to the conclusion that the driver for the strong amplification of the torsional oscillation of the roadway was not a resonance effect, but a so-called self-induced "single-degree-of-freedom torsional flutter", a well-known fluid-structure interaction effect in aeronautics (aeroelasticity). In Ref. [76] details of this kind of argumentation are presented.

[26]The Reynolds number built with these values is $Re = 3 \cdot 10^6$.

Fig. 6.33 Tacoma Narrows Bridge in a slight storm on Nov. 7, 1940. Due to flow separation on parts of the roadway leading to large oscillations, a lift up of the roadway up to 9 m had occurred, (above), and a twist of the roadway by angles of $\pm 45°$ could be observed (below), [77]

Fig. 6.34 The final collapse of the Tacoma Narrows Bridge, [77]

6.16 The Rayleigh Number

John William Strutt, third Baron Rayleigh, a British physicist, was born on Nov. 12, 1842 in Maldon, England and died on June 30, 1919 in Witham, England. From 1879 to 1884 he was the second Cavendish professor of experimental physics at Cambridge in succession to James Clark Maxwell. He was a fellow of the Royal Society (1873), served as its Secretary from 1885–1896 and as President from 1905–1908. In 1908 he accepted the post of Chancellor of the University of Cambridge. Rayleigh received a lot of honorary degrees and government awards amongst others the Nobel Prize in 1904, [78, 79]. His scientific work has encompassed the whole field of physics, in particular the wave theory, light scattering, acoustics, density of gases, electrodynamics, hydrodynamics, the analysis of dimensions, etc.. In 1871 Rayleigh published his theory of scattering (Rayleigh scattering law), where for the first time a correct explanation was given why the sky is blue.

John William Strutt, third Baron Rayleigh

The first volume of his great book, *The Theory of Sound*, appeared in print in 1877, followed by the second volume in 1878. His greatest contribution to science was the isolation of the gas argon from the atmosphere, for which he received the Nobel Prize, [80].

The Rayleigh number describes flows, which are governed by buoyancy, friction and heat conduction, [81]. Free convection flows belong to this class of problems. The Rayleigh number is the product of the Grashof number (Eq. (3.3)) and the Prandtl number (Eq. (6.27))

$$Ra = Gr \cdot Pr = \frac{g\beta l^3 \Delta T}{\nu^2} \cdot \frac{\nu}{k} = \frac{g\beta l^3 \Delta T}{\nu k} , \tag{6.39}$$

with g being the gravitational acceleration, β the volumetric coefficient of thermal expansion, l a reference length, ΔT an appropriate temperature difference, ν the kinematic viscosity and k the thermal diffusivity. This number has its fundamental significance for the free convection flows of Rayleigh–Bénard type, [82].

6.16.1 A Famous Example of Rayleigh Number Dependency

The Rayleigh–Bénard Convection

Given is a horizontal fluid layer, which is in contact with a (hot) bottom surface of temperature $T + \Delta T$ and a (cold) top surface of temperature T. The thickness of the fluid layer is L, Fig. 6.35.

The heat is transferred upwards only by conduction as long as the Rayleigh number Ra is lower than the critical Rayleigh number Ra_c. The critical Rayleigh number (Ra_c) is determined by theoretical considerations to $Ra_c = 1708$. In that case any motion of the fluid is prevented by the viscosity and the heat conductivity. When the Rayleigh number exceeds 1708, a thermal instability arises by a beginning buoyancy effect, which is the driver of the convection of heat. This leads to alternating patterns of upward and downward motions. In that case the Nusselt number Nu, Eq. (6.13), becomes larger than unity ($Nu > 1$). For free upper boundaries honeycomb patterns of hexagonal cells are generated, Figs. 6.36, 6.37.

For rigid upper boundaries horizontal rolls arise, Fig. 6.38. A further increase of the Rayleigh number leads to turbulent convection flows, where the cell structure is maintained, but shows periodic oscillations. And later the convection becomes chaotic.

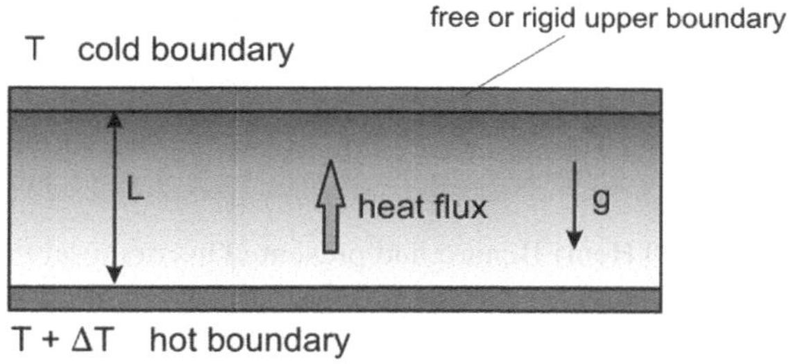

Fig. 6.35 Sketch of the Rayleigh—Bénard convection arrangement

Fig. 6.36 Rayleigh–Bénard convection: hexagonal convection pattern in a layer of silicone that is heated uniformly below and is exposed to a free upper boundary (ambient air). The fluid is rising in the cell center, moves to the edge and descends to the bottom. Photograph by Velarde, M. G., Yuste, M., and Salan, J., [27]

Fig. 6.37 Rayleigh–Bénard convection: cloud formation in the atmosphere showing regular cells. Photograph recorded by Space Shuttle Endeavor, (NASA, 1992)

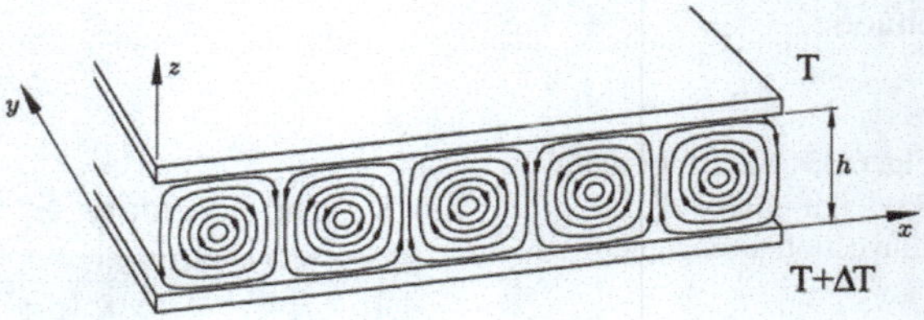

Fig. 6.38 Horizontal rolls generated by a Rayleigh–Bénard convection arrangement with a rigid upper boundary

In 1900 Henri Bénard had presented his doctoral thesis, where he reported about the observations regarding the flow effect of a fluid heated on the bottom side of a vessel, [83]. Sixteen years later J. W. Strutt, third Baron Rayleigh, gave a theoretical explanation of this effect in reference [84]. Since that time this effect is named Rayleigh–Bénard convection.

The regime of the Rayleigh number for different arrangements of the Rayleigh–Bénard convection varies over a large band. As an example, the Rayleigh number of a vessel filled with water up to an altitude of 10 cm is calculated, where between the bottom and the top a temperature difference of $\Delta T = T_b - T_t = 100\,\text{K}$ is assumed. With $\nu_{water} \approx 10^{-6}\,\text{m}^2/\text{s}$, $k_{water} \approx 1/7 \cdot 10^{-6}\,\text{m}^2/\text{s}$, $\beta = 2.07 \cdot 10^{-4}\,1/\text{K}$, $g = 9.81\,\text{m/s}$ we obtain $Ra = 1.42 \cdot 10^9$. In the atmosphere Rayleigh numbers up to 10^{20} can be achieved.

The Nusselt number, which describes the ratio of the heat transfer to the molecular heat flux ($Nu = \alpha\, l/\lambda$), is mainly a function of the Rayleigh number and the Prandtl number (see Sect. 6.9), $Nu = Nu(Ra, Pr)$. It seems that the Nusselt number describing the process of the Rayleigh–Bénard convection can be formulated by a power law

$$Nu \sim Ra^n \cdot Pr^m$$

There are a lot of experimental investigations and theoretical explanations for the determination of the coefficients n and m, [85]. Assuming moderate Prandtl numbers it was exemplarily found

$$Nu \sim Ra^{\frac{2}{7}} \text{ for } 10^6 < Ra < 10^{11} ,$$
$$Nu \sim Ra^{0.32} \text{ for } Ra = 10^{17} .$$

Other experiments resulted in $n = 0.5$ for $Ra = 10^{11}$, where the fluid was liquid helium. Further experiments have shown that these exponents can dramatically change for very low and very high Prandtl numbers, [82].

6.17 The Stokes Number

George Gabriel Stokes, 1st Baronet, was an Irish mathematician and physicist. He was born on August 13, 1819 in Skreen, Ireland, and died on February 1, 1903 in Cambridge, England. G. G. Stokes was appointed in 1849 as a professor of mathematics (the Lucasian professorship) at Cambridge University. He was the secretary of the Royal Society since 1854 and its president during the period 1885–1890. In 1889 he was made a Baronet by Queen Victoria and served as a representative of his university in the British Parliament, [86].

Sir George Gabriel Stokes

His scientific work has encompassed a variety of physical disciplines, like the theory of light (optics), acoustics and the motion as well as the inner friction of fluids. Besides other scientists, see Chap. 4, he had extended the three dimensional inviscid momentum equations of hydrodynamics (Euler equations) to flows with friction, which were named since then the Navier–Stokes equations. Further he has contributed to the mathematical physics with outcomes now known as Stokes' theorem and Stokes' law, [87].

The dimensionless number compounded by the product of the Euler number, Eq. (3.2), and the Reynolds number, Eq. (2.12)

$$Sto = Eu \cdot Re = \frac{p}{\rho u^2} \cdot \frac{\rho u l}{\mu} = \frac{pl}{\mu u} \quad \text{Stokes number} , \tag{6.40}$$

bears Stokes' name, [4, 81].

6.17.1 The Significance of the Stokes Number in the Light of The Navier–Stokes Equations

The steady, incompressible, two dimensional momentum equations read

$$u\frac{\partial u}{\partial x}+v\frac{\partial u}{\partial y}=-\frac{1}{\rho}\frac{\partial p}{\partial x}+\nu\left(\frac{\partial^2 u}{\partial x^2}+\frac{\partial^2 u}{\partial y^2}\right),$$
$$u\frac{\partial v}{\partial x}+v\frac{\partial v}{\partial y}=-\frac{1}{\rho}\frac{\partial p}{\partial y}+\nu\left(\frac{\partial^2 v}{\partial x^2}+\frac{\partial^2 v}{\partial y^2}\right). \quad (6.41)$$

We make these equations dimensionless by

$$x'=\frac{x}{l_x},\quad y'=\frac{y}{l_y}.\quad u'=\frac{u}{u_\infty},\quad v'=\frac{l_x}{l_y}\frac{v}{u_\infty},\quad p'=\frac{p}{\Delta p},$$

and assume further

$$x'\approx O(1),\quad y'\approx O(1),\quad u'\approx O(1),\quad v'\approx O(1),\quad p'\approx O(1).$$

For flows with small Reynolds numbers ($Re << 1$)—creeping flows—the inertial terms can be neglected and one obtains

$$\underbrace{Re\left(u'\frac{\partial u'}{\partial x'}+v'\frac{\partial u'}{\partial y'}\right)}_{<<1}-\left(\frac{\partial^2 u'}{\partial x'^2}+\frac{\partial^2 u'}{\partial y'^2}\left(\frac{l_x}{l_y}\right)^2\right)=-Sto\frac{\partial p'}{\partial x'},$$
$$\underbrace{Re\left(\frac{l_y}{l_x}\right)^2\left(u'\frac{\partial v'}{\partial x'}+v'\frac{\partial v'}{\partial y'}\right)}_{<<1}-\left(\frac{l_y}{l_x}\right)^2\left(\frac{\partial^2 v'}{\partial x'^2}+\frac{\partial^2 v'}{\partial y'^2}\left(\frac{l_x}{l_y}\right)^2\right)=-Sto\frac{\partial p'}{\partial y'}, \quad (6.42)$$

with $Sto = Eu \cdot Re$. Equation (6.42) are the Stokes equations.

A First Characteristic Example of the Stokes Number

For the flow case of a slide bearing with the longitudinal dimension l_x much larger than the lateral dimension l_y and $(l_y/l_x)^2 << 1$, Eq. (6.42) reduces to (Fig. 6.39)

$$Sto\frac{\partial p'}{\partial x'}=\left(\frac{l_x}{l_y}\right)^2\frac{\partial^2 u'}{\partial y'^2},$$
$$Sto\frac{\partial p'}{\partial y'}=\frac{\partial^2 v'}{\partial y'^2}. \quad (6.43)$$

Comparison of the orders of magnitude in Eq. (6.43) reveals that

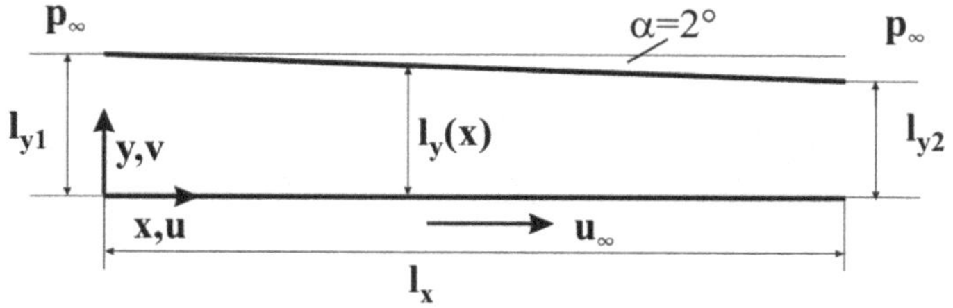

Fig. 6.39 Sketch of a slide bearing

$$\frac{\partial p'}{\partial x'} >> \frac{\partial p'}{\partial y'} ,$$

which means that the influence of the pressure in y-direction can be neglected. Integration of dp/dx with respect to the boundary conditions of the slide bearing, $y = 0,\ u = u_\infty,\ x = 0,\ p = p_\infty$ and $y = l_y(x),\ u = 0,\ x = l_x,\ p = p_\infty$, (Fig. 6.13), yields, [88]

$$\begin{aligned} p &= p_\infty + 6\mu u_\infty \frac{1}{l_x} F(l_y) , \\ Sto &= Eu \cdot Re = 6F(l_y) , \end{aligned} \tag{6.44}$$

with[27]
$$F(l_y) = \frac{l_x^2}{(l_{y1}^2 - l_{y2}^2)} \frac{(l_{y1} - l_y)(l_y - l_{y2})}{l_y^2} .$$

As an example for the slide bearing flow we have chosen: $l_x = l = 1,\ l_{y1} = la,\ l_{y2} = l(a - \tan\alpha),\ l_y = l_{y1} - x \tan\alpha,\ a = 0.2,\ \alpha = 2°$. Evaluation of Eq. (6.44) results in a Stokes number distribution which is in agreement with the creeping flow estimations, $Sto = O(1)$, Fig. 6.40.

A Second Characteristic Example of the Stokes Number

In 1851 G. G. Stokes presented an analytical solution for the creeping flow ($Re << 1$) past a sphere (Solution of the linear Stokes equations in spherical coordinates.). Thus the pressure distribution is given to, [89, 90]

$$p = p_\infty - \frac{3}{2}\mu u_\infty \frac{a \cos\Theta}{R^2} , \tag{6.45}$$

with a being the radius of the sphere, Θ the azimuth angle of the spherical coordinates and R the spherical distance of a point in the flow field, Fig. 6.41. In the sphere flow the dimensions of l_x and l_y are of the same order. Therefore we find from Eq. (6.42) the order of magnitudes

[27] We have used here the Euler number in the form $Eu = \dfrac{\Delta p}{\rho u^2}$.

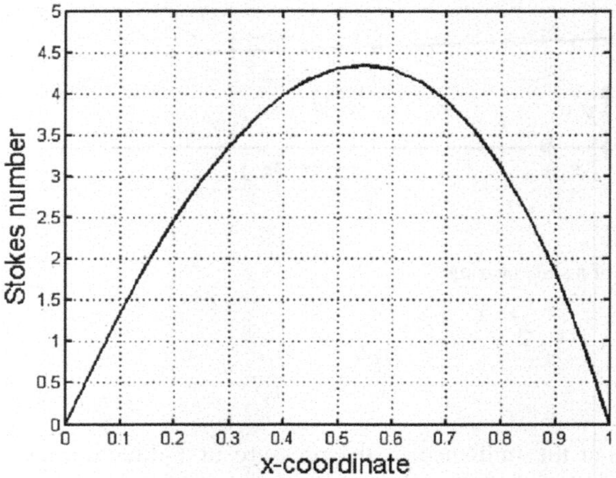

Fig. 6.40 Stokes number computation evaluating Eq. (6.44)

$$\frac{l_x^2}{l_y^2} = O(1), \quad Sto = O(1)\,. \tag{6.46}$$

The equation of the pressure (Eq. (6.45)) can be reformulated to

$$Eu = -3\frac{1}{Re}\frac{a^2}{R^2}\cos\Theta\,, \tag{6.47}$$

with $Re = \frac{u_\infty\, 2a}{\nu}$, and on the surface of the sphere ($R = a$) we obtain

$$Sto = Eu \cdot Re = -3\cos\Theta\,, \tag{6.48}$$

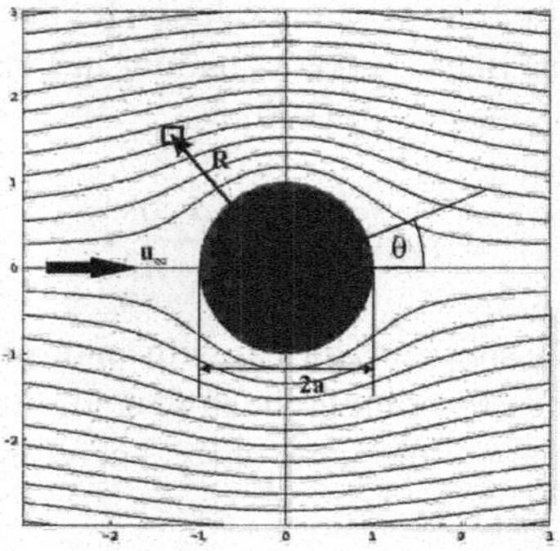

Fig. 6.41 Sketch of a creeping flow around a sphere

Fig. 6.42 Demonstration of a creeping flow: Hele-Shaw flow past a circle. Photograph by D. H. Peregrine, [27]

which satisfies the estimations of Eq. (6.46), [4]. Figure 6.42 demonstrates the creeping flow (Stokes flow) past a circle positioned between two parallel flat plates with very small gap width (Hele-Shaw flow) .

Another often used definition of the Stokes number, in particular for optical measurement systems in fluid dynamics (PIV ⇒ particle image velocimetry) is

$$Sto = \frac{t_0 u_0}{l_0}\,, \tag{6.49}$$

where t_0 is the relaxation time of a particle, u_0 denotes the velocity of the fluid and l_0 is a typical length of the fluid (e.g.: the diameter of an obstacle or object in the fluid). For Stokes flows ($Re << 1$) t_0 can be defined by

$$t_0 = \frac{\rho_p d_p^2}{18\mu_f}\,, \tag{6.50}$$

with ρ_p and d_p being the particle density and diameter, and μ_f the dynamic viscosity of the fluid. The relaxation time t_0 is the time required for particles to respond to changes of the fluid velocity such that when the surrounding fluid velocity varies, as it might when encountering an object, the particles follow the flow around that object. Stokes numbers much larger than unity, $Sto >> 1$, describe particles which remain unaffected by a fluid velocity change and continue their original trajectories. For Stokes numbers smaller than unity, $Sto << 1$, the particles follow the fluid velocity and the streamlines, [91].

This behavior is clearly illustrated by the sketch, taken from [92], Fig. 6.43. There the flow with entrained particles through a cover plate is drawn and shows the trace of a streamline of the gas and the trajectory of a particle. For small Stokes numbers ($Sto << 1$) the trajectory of the particle follows more or less directly the streamline, but for large Stokes numbers ($Sto >> 1$) the inertia of a particle prevails and its trajectory proceeds as shown in the figure, Fig. 6.43.

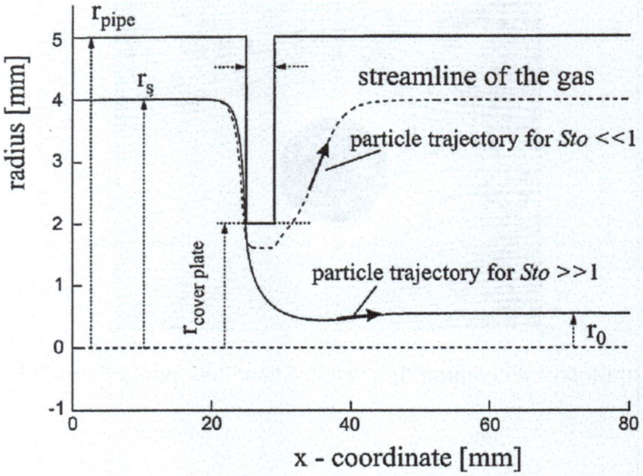

Fig. 6.43 Flow through a cover plate. Gas with entrained particles. Sketch taken from [92]

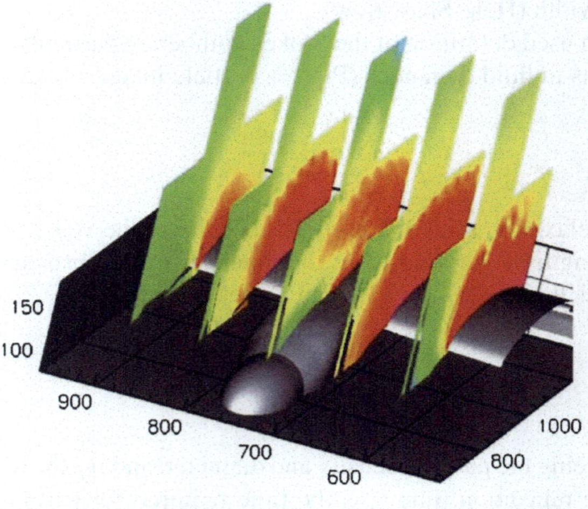

Fig. 6.44 Flow field measurement (velocity magnitude) around an airplane model using the particle image velocimetry (PIV) method, [93]

In Fig. 6.44 the results of velocity distributions for a flow field past an aircraft model are shown, which are achieved by the experimental particle image velocimetry (PIV) method,[28] [93]. The slippage of the injected particles with respect to the streamlines is object of calibrating test measurements, where in general $Sto << 1$ is required. For a Stokes number $Sto \approx 0.4$ an uncertainty of the measurement data of order 10 percent has to be expected.

[28] Also known as Laser-Doppler-Anemometry.

6.18 The Knudsen Number

Martin Hans Christian Knudsen

Martin Hans Christian Knudsen was a Danish physicist. He was born on Feb. 15, 1871 in Hasmark and died on May 27, 1949 in Copenhagen, Denmark. He had his study of physics at the University of Copenhagen, where he graduated in 1896. In 1912 he accepted a post as professor at this university, where he remained until his retirement in 1941. During the period 1927–1928 he act as rector of the university. His scientific work has encompassed two main interests, the kinetic theory of gases at low pressures (rarefied gases, molecular flows) and the hydrography, where he was the Danish delegate in the "Central Committee for Oceanic Research of the International Council for Exploration of the Sea" for the long period 1902–1947, [94]. Today we know from him the Knudsen effect, the Knudsen gas, the Knudsen cell etc. and in particular the Knudsen number, which characterizes the various flow regimes from continuum to free molecular flow, [24].

The definition of the Knudsen number is given by the ratio of the free mean path[29] λ of the flow molecules to a characteristic length L of the flow considered:

$$Kn = \lambda/L \,. \tag{6.51}$$

6.18.1 Knudsen Number Relations

The definition of the Knudsen number can be reformulated in a remarkable manner. Using the findings of the theory of molecular gasdynamics for the description of the dynamic viscosity μ and the free mean path λ

$$\mu = \frac{5\sqrt{\pi k m T}}{16\pi\sigma^2} \quad \text{and} \quad \lambda = \frac{m}{\sqrt{2}\pi\sigma^2\rho}\,, \tag{6.52}$$

with k being the Boltzmann constant ($k = 1.38 \cdot 10^{-23}$ Nm/K), m the mass of a molecule ($m = 4.8 \cdot 10^{-23}$ g for air), σ the diameter of a molecule ($\sigma = 3.7 \cdot 10^{-10}$ m for air), ρ the density ($\rho = 1.292$ kg/m^3 for air) and T the temperature, one finds

$$\lambda \approxeq \frac{\mu}{\rho}\sqrt{\frac{\pi m}{2kT}} \quad \text{with} \quad \frac{16}{5\pi} \approxeq 1.0\,. \tag{6.53}$$

[29] We use here the Greek letter λ to denote the free mean path, since it is common in the literature, knowing that λ is also used for the heat conduction coefficient.

Then the following interesting relation can be derived [95, 96]

$$Kn \approx \frac{1}{L}\frac{\mu}{\rho}\sqrt{\frac{\pi m}{2kT}} = \underbrace{\frac{\mu}{\rho u L}}_{1/Re_L} \cdot \underbrace{\frac{u}{a}}_{M} \cdot a\sqrt{\frac{\pi m}{2kT}} = \frac{M}{Re_L}\sqrt{\gamma}\sqrt{\pi/2}\,,$$

$$Kn \approx \frac{M}{Re_L} 1.28\,\sqrt{\gamma}\,, \tag{6.54}$$

with Re_L being the Reynolds number, M the Mach number, $a = \sqrt{\gamma}\sqrt{\frac{\pi}{2}}\sqrt{\frac{2kT}{\pi m}}$ the speed of sound and γ the ratio of specific heats.

For a perfect gas, which satisfies the Maxwell–Boltzmann distribution, one obtains from the theory of gas kinetics

$$Kn = \frac{kT}{\sqrt{2}\pi\sigma^2\, p} \cdot \frac{1}{L}\,. \tag{6.55}$$

which relates with $\rho = n\,\mathrm{m}$, $p = n\,kT$ ($n = \rho/m = 0.269 \cdot 10^{26}\ 1/m^3$ the particle density) and Eq. (6.52) also to

$$Kn = \lambda/L\,.$$

In the case that the Knudsen number is built with the laminar boundary layer thickness δ

$$Kn_\delta = \frac{\lambda}{\delta}\,,$$

one obtains owing to $\frac{\delta}{L} \sim \frac{1}{\sqrt{Re_L}}$ (Eq. 4.28) for laminar boundary layer flows [24, 97]

$$Kn_\delta \sim \frac{M}{\sqrt{Re_L}}\,. \tag{6.56}$$

6.18.2 Knudsen Numbers in Hypersonic Flows

The mission of typical space vehicles, in particular when manned, consists of an ascent from Earth ground, an operational phase in space, when travelling to the moon or other planets or moving in Earth orbits, and finally a re-entry phase in Earth' atmosphere. The most critical part is the re-entry process, where in the beginning the space vehicle is exposed to Mach numbers $M > 30$ in the upper atmosphere,

Table 6.6 List of flow regimes characterized by the Knudsen number for a typical re-entry flight situation [24, 96]

Continuum flow	$0 < Kn < 10^{-2}$
Continuum flow with slip effects (slip flow and temperature jump)	$10^{-2} < Kn < 10^{-1}$
Disturbed molecular flow (rarefied gas) (particles collide with each other and the wall)	$10^{-1} < Kn < 10^{1}$
Free molecular flow (particles collide only with the wall)	$10^{1} < Kn$

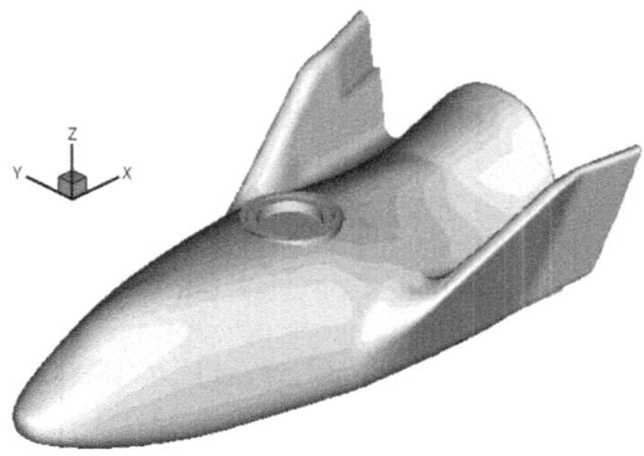

Fig. 6.45 X-38 vehicle, [38]

which decelerates along a given trajectory to zero. During the path along the flight trajectory the space vehicle crosses various flow regimes, which are characterized by the Knudsen number, Table 6.6.

The space flight vehicle X-38, Fig. 6.45, was a common project between NASA[30] (USA) and ESA[31] (Europe) in the 1990s, to act as a Crew Rescue Vehicle (CRV) for the crew members of the International Space Station (ISS). Unfortunately, in 2002 NASA changed its strategy and the X-38 project was cancelled. Up to now (2020) no winged follow-on project was launched. Nevertheless, the work on the aerodynamic data base was brought to a reasonable end. Figure 6.46 gives the flight trajectory of the X-38 vehicle, the Knudsen number distribution and the flow regimes.

[30]National Aeronautics and Space Administration.

[31]European Space Agency.

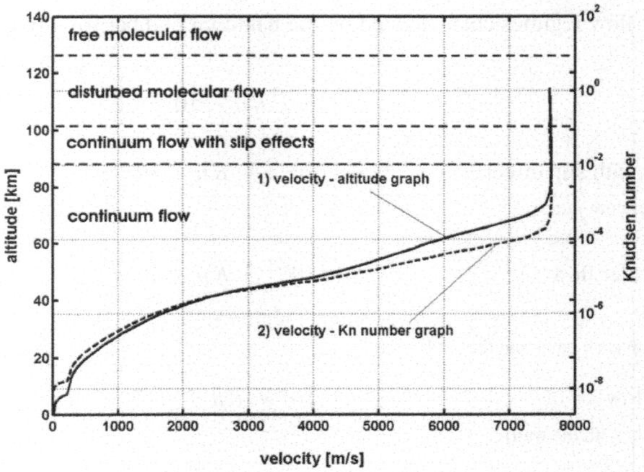

Fig. 6.46 X-38 trajectory: presented as (1) velocity-altitude graph and (2) velocity-Knudsen number graph with an approximate reference length for the X-38 vehicle of $L = 8$ m, [37]

6.19 The Mach Number

Ernst Mach was a Czech,—at that time part of the Austro-Hungarian Monarchy—, scientist with interests in mathematics, physics, philosophy and later also in physiology of sensory perceptions. He was born in Chrlice (Czech Republic) on Feb. 18, 1838 and died on Feb. 19, 1916 in Haar, Germany. After a short time as professor in Graz he accepted 1867 a post as professor for experimental physics at the Charles University of Prague, where he remained until 1895. Then he received a professorship for inductive philosophy at the University of Vienna, Austria, and retired from this post in 1901, when he was appointed to the Austrian parliament. His scientific work has covered a lot of aspects of the physics of optical waves, like interference, diffraction, refraction, polarization, etc.. In the field of gasdynamic his work regarding the flow structure of projectiles moving with supersonic speeds (—first identification of shock waves—) was of seminal importance until this day, [98, 99]. The ratio of the flight velocity u of a flight object to the speed of sound a bears his name[32] in honor of his fundamental contribution to the understanding of supersonic flows:

Ernst Mach

[32] J. Ackeret, 1898–1981, obviously was the first, who in his habilitation treatise has named this ratio Mach number, [100].

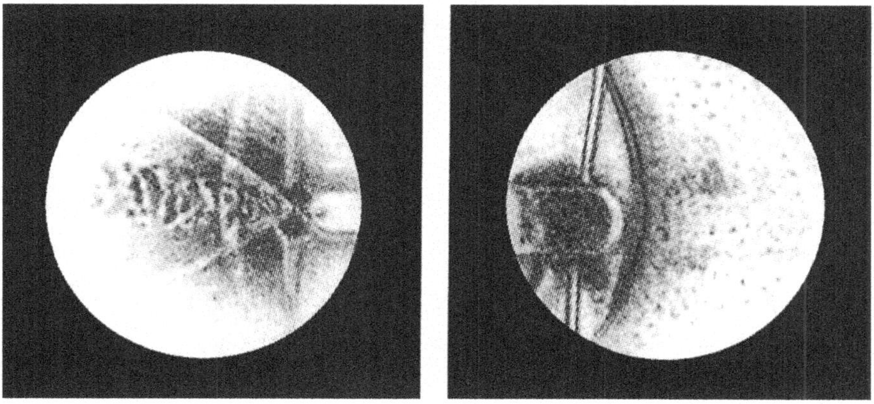

Fig. 6.47 Presumably the first photographs of shockwaves, recorded by P. Salcher and E. Mach, [101, 102]. Schlieren photographs of shockwaves at the tail (left) and the tip (right) of a projectile

$$M = \frac{u}{a}$$

In Eqs. (4.19), (6.1), and (6.9) we encountered already the Mach number. At least Eq. (6.1) was known long before Ernst Mach has conducted his gasdynamic studies, which means that the relation now known as the Mach number was implicitly present already since Leonhard Euler. Figure 6.47 shows probably the first photographs of shock waves,[33] [101, 102].

6.19.1 *Some Impressions Regarding the Relevance of the Mach Number*

The most important dimensionless numbers in aerodynamics are the Reynolds number Re and the Mach number M. In aerothermodynamics (hypersonic and space flight) a third significant dimensionless number, the Prandtl number Pr, has to be added.

The flight of civil and military aircraft takes place in different altitudes of our atmosphere. The temperature in the atmosphere depends on the altitude and with that also the speed of sound ($a = \sqrt{\gamma R T}$). In Table 6.7 we have computed for a constant Mach number ($M = 0.85$) the speed of the flight object as function of the altitude, in order illustrate that a constant Mach number does not mean a constant velocity. An even stronger situation is encountered in the European Transonic Wind

[33] In [98] is reported that E. Mach has believed, when he saw the photographs at the first time, that air mass was compressed in the front part (tip) of the projectile. In contrary to that P. Salcher shall have argued, that it looks more than a bow wave of a ship, which was obviously the right interpretation.

Table 6.7 Flight velocities at different altitudes for a constant Mach number of $M = 0.85$

u in [m/s] for $M = 0.85$	Altitude H in km	Speed of sound a in m/s	Temperature T in K
289.24	0	340.29	288.15
254.60	10	299.53	223.25
250.80	20	295.07	216.65
256.44	30	301.70	226.50
269.61	40	317.19	250.35

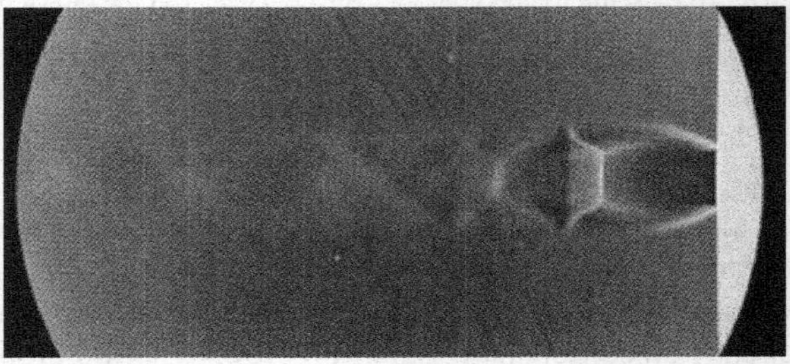

Fig. 6.48 Photograph of air jet flow performed by E. Mach and P. Salcher, [103]

Tunnel (ETW), where test temperatures of 90 K are possible, see also Sect. 6.5. The speed of sound under sea level conditions amounts to 340.29 m/s and for the ETW test conditions at 90 K 190.18 m/s, which for $M = 1$ are also the freestream velocity values. In other words at $M = 1$ the velocity of a flight vehicle at sea level conditions is 150.11 m/s larger than in the ETW.

In the 1890s E. Mach and P. Salcher had also conducted experiments with supersonic jet flows, Fig. 6.48, where they tried to improve their optical experience with the Schlieren photography, [103]. At that time it was difficult to give a comprehensive explanation about what they saw. Today we have a complete understanding of such kind of flows, Fig. 6.49, which consists of a complicated conglomerate of oblique and normal shocks, shear layers, triple points and all sorts of reflections at the fluid boundaries, [104].

An impressive picture is given by Fig. 6.50, where due to the transonic flight Mach number (M close to unity) a vapor surface was formed by condensation of the water vapor of the air.

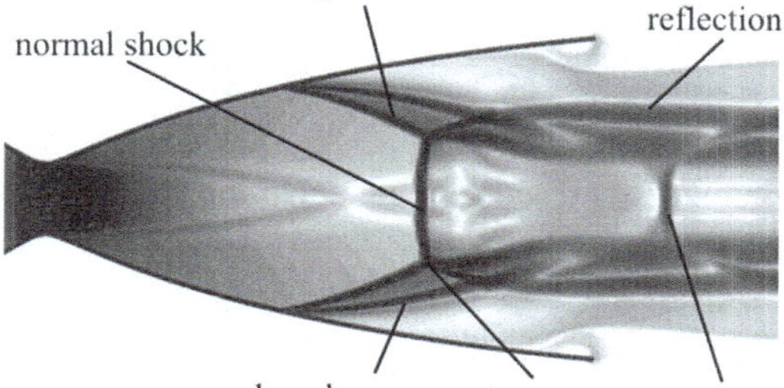

Fig. 6.49 Flow in an expansion nozzle showing all details of a supersonic jet flow. Pseudo Schlieren taken from a Navier–Stokes solution, [104]

Fig. 6.50 Combat aircraft at transonic flight, where the Mach number is close to unity. Due to the supersonic expansion a vapor surface just before the terminating shock is formed owing to condensation

6.20 The Stefan Number

Josef Stefan was an Austrian mathematician and physicist. He was born on March 24, 1835 in St. Peter (Klagenfurt), Austria and died on Jan. 7, 1893 in Vienna, Austria. J. Stefan had his study at the University of Vienna and graduated in mathematical physics in 1857. In 1863 he became a professor at this University and later (1866) he was appointed to be the director of the Physical Institute, [105]. His scientific work on the one hand has dealt with the determination of the conductivity of gases. On the other hand he showed experimentally in 1879, that the total radiation from a blackbody was proportional to the fourth power of the absolute temperature. This was his greatest contribution to the physics. Later in 1884 his student Ludwig Boltzmann gave a theoretical explanation of the physical behavior. Therefore, today, we speak about the Stefan–Boltzmann law.

Josef Stefan

The Stefan number is defined by Eq. (3.9)

$$Ste = \frac{\varepsilon \sigma T^3 l}{\lambda} . \tag{6.57}$$

with ε being the emission coefficient, σ the Stefan–Boltzmann constant, λ the thermal conductivity, T the temperature and l a characteristic length of the flow considered. The use of the Stefan number is not very widespread. Unfortunately, there are two definitions of the Stefan number. The first one is that of Eq. (6.57), which is applicable for hypersonic flows past flight objects (e.g. space vehicles), where due to strong heat transfer effects hot vehicle surfaces exist. Heat is radiated from these surfaces, and additionally the gas itself radiates, when it stands in a highly excited non-equilibrium thermodynamic state. This kind of Stefan number is sometimes also named Stark number or Stefan–Boltzmann number .

The second one is defined by $Ste = c_{ps} (T_f - T_0)/h$ which denotes the ratio between sensible heat and latent heat and is to be applied for phase transition problems, like melting or solidification, [106]. The term $c_{ps}(T_f - T_0)$ is the energy removed from a solid at the fusion temperature T_f to lower its temperature to T_0. h is the energy, which is added during phase transformation (e.g. melting) at the fusion temperature.

6.21 Other Dimensionless Numbers with Relevance to Fluid Mechanics and Heat Transfer

In the following we list three dimensionless numbers, which do not bear a name of a scientist or an engineer, but are used more or less often in gasdynamics and high Mach number flows.

Wall Temperature Parameter

A dimensionless number is also the ratio of temperatures

$$\frac{T_w - T_\infty}{T_\infty} = \frac{g l^3 (T_w - T_\infty)}{\nu^2 T_\infty} \cdot \frac{u^2}{g l} \cdot \frac{\nu^2}{u^2 l^2} = \frac{Gr \cdot Fr}{Re^2} , \tag{6.58}$$

with the Reynolds number Re, the Froude number Fr, Eq. (3.2), and the Grashof number Gr, Eq. (3.3) as well as the wall temperature T_w and the reference temperature T_∞. This parameter is important for hypersonic flows, where the surface of a flight vehicle is heated up through heat transfer processes (convection, conduction, dissipation, radiation and thermal surface effects), [24].

Binary Scaling Parameter

The binary scaling parameter

$$\rho\, l$$

is important for flows with dissociation, where ρ denotes the density of the gas and l the characteristic length of a flight vehicle and comes into play for model tests in high-enthalpy facilities. The binary scaling parameter is not a dimensionless number.

Ratio of Specific Heats

Finally, also the ratio of specific heats of gases

$$\gamma = \frac{c_p}{c_v}\,, \tag{6.59}$$

is a dimensionless number.

6.22 Dimensionless Numbers, which are not Often Used

In this section we address dimensionless numbers, which are applied only in very specific flow and/or heat transfer cases.

6.22.1 The Archimedes Number

The Archimedes[34] number is defined by

$$Ar = \frac{g\, l^3}{\mu^2}\, \rho_F(\rho_B - \rho_F), \tag{6.60}$$

with g being the gravitational acceleration, l a reference length, μ the dynamic viscosity of the fluid, ρ_F the density of the fluid and ρ_B the density of the solid body. It describes the ratio of the lift force to the friction force. The equation Eq. (6.60) looks formally similar to the Grashof number, Sect. 6.4, but ρ_F is replaced by ρ_w, the density at the wall, and ρ_B by ρ_∞, the free-stream density.

Let us consider a spherical particle of aluminium with diameter d sinking in a vessel filled with water, Fig. 6.51. The particle has a constant sinking velocity u, when the acting forces are in equilibrium. Further we assume Stokes flow ($Re < 1$) where the drag force for a sphere is given by $F_D = 3\pi\mu d u$, hence

[34] Archimedes, the famous Greek universal genius, about 287–212 B.C.

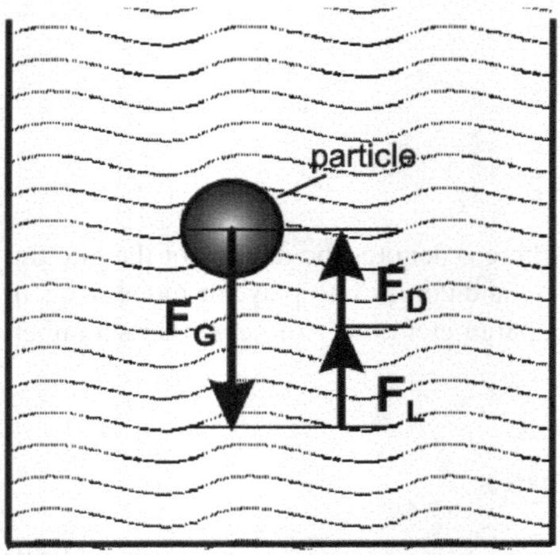

$\rho_F = 1 \cdot 10^3\ kg/m^3$ density of water,
$\rho_B = 2.7 \cdot 10^3\ kg/m^3$ density of aluminium,
$\nu = 1 \cdot 10^{-6}\ m^2/s$ kinematic viscosity of water,
$g = 9.81\ m/s^2$ gravitational acceleration.

Fig. 6.51 Sink rate u of a spherical particle of aluminum in water for Stokes flow

$$F_G - F_L - F_D = 0\,, \tag{6.61}$$

with

$$F_L = \frac{\pi g d^3}{6}\rho_F \quad \text{lift force}\,,$$
$$F_G = \frac{\pi g d^3}{6}\rho_B \quad \text{gravitational force}\,. \tag{6.62}$$

From Eq. (6.61) we find for the sink rate u

$$u = \frac{d^2 g(\rho_B - \rho_F)}{18\mu} = \frac{\nu}{18d}\cdot\frac{d^3 g(\rho_B - \rho_F)}{\nu^2\rho_F} = \frac{\nu}{18d}\cdot Ar\,, \tag{6.63}$$

Stokes flow appears only for very small sphere diameters. If we assume $d = 10^{-4}$ m we find a sink rate of $u = 0.9265 \cdot 10^{-2}$ m/s for a Reynolds number $Re = 0.9265$ and an Archimedes number $Ar = 16.67$.

6.22.2 The Bingham Number

The Bingham[35] number is defined by

[35]E. C. Bingham, an US chemist (1878–1945).

$$Bm = \frac{\tau_y l}{\mu u}, \tag{6.64}$$

with τ_y being the yield stress, l a reference length, μ the dynamic viscosity and u the flow velocity. This number denotes the ratio of yield stress to viscous stress and is used to study the flow of Bingham fluids (non-newtonian fluids, rheology), [107].

6.22.3 The Biot Number

When we consider the heat transfer between the surface and the bulk conductivity of solids, the Biot[36] number comes into play.

$$Bi = \frac{\alpha l}{\lambda_s}, \tag{6.65}$$

with α being the heat transfer coefficient (fluid to wall), l a reference length,[37] and λ_s the thermal conductivity of the solid. The Biot number describes the ratio between the heat transferred to the wall (of a solid) to the thermal conductivity inside the solid,[38] [18]. Small Biot number ($Bi < 0.1$) means that large conductivity inside the solid exists, leading to small temperature gradients in the solid.

6.22.4 The Bodenstein Number

The product of the Reynolds number, Eq. (2.12), and the Schmidt number, Eq. (6.26), yields the Bodenstein[39] number Bo:

$$Bo = \frac{ul}{D}, \tag{6.66}$$

with u being the velocity, l a reference length and D the mass diffusion coefficient. This number describes the convective to the mass diffusive transport rates. In chemical engineering processes (reactors) often the axial diffusion coefficient D_{ax} is used in the definition of the Bodenstein number.

[36] J.-B. Biot, a French physicist (1774–1862).

[37] The reference length is mostly defined by the ratio between the volume of the body to the surface of the body $l = V_{body}/S_{body}$.

[38] The definition of the Nusselt number, Eq. (2.39), is formally the same as the Biot number, but in the Nusselt number definition the thermal conductivity is that of the fluid and in the Biot number definition that of the solid.

[39] M. E. A. Bodenstein, a German physicist (1871–1942).

6.22.5 The Brinkman Number

Multiplying the Prandtl number Eq. (6.27) with the Eckert number Eq. (6.10) results in the Brinkman[40] number Br:

$$Br = Pr \cdot \tilde{Ec} = \frac{\mu u^2}{\lambda(T_w - T_{ref})}, \tag{6.67}$$

with μ being the dynamic viscosity, u the velocity, λ the heat conduction coefficient, T_w the wall temperature and T_{ref} a reference temperature of the fluid.

The Brinkman number is the ratio between the heat generated by viscous dissipation to heat transported by conduction. This number is used in chemical engineering for polymer processing, [108].

6.22.6 The Galilei Number

When the square of the Reynolds number, Eq. (2.12), is divided by the Froude number, Eq. (6.2), one gets the Galilei[41] number Ga:

$$Ga = \frac{Re^2}{Fr} = \frac{u^2 l^2}{\nu^2} \cdot \frac{gl}{u^2} = \frac{gl^3}{\nu^2}, \tag{6.68}$$

with g being the gravitational acceleration, l a reference length and ν the kinematic viscosity. This number characterizes the ratio of the gravity force to the molecular momentum transfer, [109, 110]. Another relation shows the connection of the Galilei number with the Grashof number Gr, Sect. 6.4. If we divide the Grashof number by the dimensionless relation $(\rho_w - \rho_\infty)/\rho_\infty$ we find, see also Sect. 6.22.1,

$$Ga = \frac{Gr \cdot \rho_\infty}{(\rho_w - \rho_\infty)} = \frac{gl^3}{\nu^2}. \tag{6.69}$$

6.22.7 The Laplace Number

The Galilei number can be found by dividing the square of the Reynolds number by the Froude number, see Eq. (6.68). When the square of the Reynolds number, instead of the Froude number, is divided by the Weber number, Eq. (6.4), the Laplace[42] number is obtained.

---

[40] H. C. Brinkman, a Dutch physicist (1908–1961).

[41] G. Galilei, the famous Italian astronomer (1564–1642).

[42] P.-S. Laplace, the famous French mathematician and physicist (1749–1827).

$$La = \frac{Re^2}{We} = \frac{u^2 l^2}{\nu^2} \cdot \frac{\rho u^2 l}{\sigma} = \frac{\sigma \rho l}{\mu^2}\,, \tag{6.70}$$

where σ denotes the surface tension, ρ the density, l a reference length and μ the dynamic viscosity. The fluid dynamics of free surface flows can be characterized by the Laplace number, which describes the ratio of surface tension to the molecular momentum transfer.

6.22.8 The Ohnesorge Number

The Ohnesorge[43] number has the form

$$Oh = \frac{\mu}{\sqrt{\rho \sigma l}} = \frac{\sqrt{We}}{Re}\,, \tag{6.71}$$

and describes the ratio between the molecular momentum transfer and the surface tension. With Eq. (6.70) it follows

$$Oh = \frac{1}{\sqrt{La}}\,.$$

The Ohnesorge number has its importance for spray generation processes and decay phases of droplets, like the Weber number has, see Sect. 6.3.

6.22.9 The Richardson Number

The inverse of the Froude number, Eq. (6.28), is the Richardson[44] number,

$$Ri = \frac{1}{Fr} = \frac{gl}{u^2}\,.$$

The Richardson number and its variants[45] are used in weather prediction problems, [111].

[43] W. Ohnesorge, a German engineer (1901–1976).

[44] L. F. Richardson, a British mathematician and meteorologist (1881–1953).

[45] In meteorology some variants of the Richardson number are used, which are denoted by *flux* Richardson number, *gradient* Richardson number, *bulk* Richardson number. For example, the *gradient* Richardson number characterizes the dynamic stability of a meteorological flow and reads $Ri_g = \frac{g}{T_v} \cdot \frac{\partial T_v}{\partial z} / \left(\left(\frac{\partial U}{\partial z}\right)^2 + \left(\frac{\partial V}{\partial z}\right)^2 \right)$.

6.22.10 The Roshko Number

The Roshko[46] number

$$Ro = Str \cdot Re = \frac{fl}{u} \cdot \frac{ul}{\nu} = \frac{fl^2}{\nu}, \tag{6.72}$$

describes oscillating fluid mechanical problems. It consists of the product of the Strouhal number Str, Eq. (6.36), and the Reynolds number Re, Eq. (2.12), where f is a characteristic frequency of a flow problem, e.g. a vortex system, shedded from a bluff body, l a reference length and ν the kinematic viscosity of the fluid, [112].

6.22.11 The Sherwood Number

In analogy to the Nusselt number, Eq. (6.13), which represents the heat transfer due to the thermal boundary layer, the Sherwood[47] number deals with the mass transfer due to the concentration boundary layer and reads

$$Sh = \frac{\beta l}{D}, \tag{6.73}$$

with β being the convective mass transfer coefficient, l a reference length and D the coefficient of mass diffusion. It describes the ratio of the convective mass transfer to the mass transferred due to diffusion. Some relations of the Sherwood number regarding flat plate flows with different flow conditions can be found in [113].

References

1. Calinger, R.S.: Mathematical Genius. Mathematical Genius in the Enlightenment. Princeton University Press, Princeton (2015)
2. Gautschi, W.: Leonhard Euler: His Life, the Man, and His Work. SIAM Rev. **50**(1), 3–33 (2008)
3. Sommerfeld, A.: Mechanik der deformierbaren Medien, Akademische Verlagsgesellschaft Geest & Portig (1954)
4. Zierep, J.: Ähnlichkeitsgesetze und Modellregeln der Strömungsmechanik. G. Braun Verlag, Karlsruhe (1972)
5. N.N.: Grace's Guide to British Industrial History: William Froude. https://www.gracesguide.co.uk/William_Froude
6. Weber, M.: In: Jahrbuch der Schiffbautechnischen Gesellschaft (1919)
7. Herwig, H.: Strömungsmechanik A - Z. Vieweg Verlag, Wiesbaden (2001)
8. Krüger, S.: Vorlesungsmanuskript. Universität Hamburg-Harburg (2009)

[46] A. Roshko, an US physicist and engineer (1923–2017).

[47] T. K. Sherwood, an US engineer of chemistry (1903–1976).

9. Saunders, H.E.: Hydrodynamics in Ship Design, vol. 2 (1957)
10. Weber, M.: Jahrbuch der Schiffbautechnischen Gesellschaft, 274–388 (1930)
11. Hüttl, C.: Einfluß der Sprayausbreitung und Gemischbildung auf die Verbrennung von Biodieselgemischen. In: Brüggemann, D. (ed.) Thermodynamik, vol. 19. Logos Verlag, Berlin (2011)
12. Pfeifer, C.: Experimentelle Untersuchungen von Einflußfaktoren auf die Selbstzündung von gasförmigen und flüssigen Brennstofffreistrahlen. Karlsruhe Institute of Technology, Scientific Publishing, Scientific Report 7555 (2010)
13. Ranz, W.E.: On Sprays and Spraying: A Survey of Spray Technology for Research and Development Engineers. Pennsylvania State Univ. USA, Bull. **1956**, 655 (1956)
14. Lin, S.P., Reitz, R.D.: Drop and Spray Formation from a Liquid Jet. Annual Review of Fluid Mechanics **30**, 85–105 (1998)
15. Levebvre, A.H.: Gas Turbine Combustion. Purdue University, Indiana, USA, Thermal and Combustion Center (1998)
16. Wauer, J.: Die Mechanik und ihre Fachvertreter an der Universität Karlsruhe. KIT Scientific Publishing, Karlsruhe (2017)
17. Böckh, V., P., Wenzel, T.: Wärmeübertragung. Springer Vieweg, Berlin (2014)
18. Polifke, W., Kopitz, J.: Wärmeübertragung Grundlagen, analytische und numerische Methoden. Pearson Studium, Munich (2009)
19. Jackson, J.D.: Osborne Reynolds - Scientist, Engineer and Pioneer. Proc. Roy. Soc. Lond. A, **451**, 49–86 (1995)
20. Fairclough, C.: Happy Birthday, Osborne Reynolds (2017). https://www.comsol.com/blogs/happy-birthday-osborne-reynolds/
21. Reynolds, O.: An experimental investigation of the circumstances which determine wether the motion of water shall be direct or sinuous, and of the law of resistance in parallel channels. R. Soc., Phil. Trans. (1883)
22. Reynolds, O.: On the dynamical theory of incompressible viscous fluids and the determination of the criterion. R. Soc., Phil. Trans. (1895)
23. Oswatitsch, K.: Österreich. Ing. Zeitschrift **6**, 421–426 (1963)
24. Hirschel, E.H.: Basics of Aerothermodynamics, 2nd edn. Springer, Berlin (2015)
25. van der Bliek, J.A.: The European Transonic Wind Tunnel. Buch und Offsetdruck Thierbach (1996)
26. www.dlr.de/mp/Portaldata/22/Resources//9_Turbulenz_in_Fluessigkeiten.pdf
27. van Dyke, M.: An Album of Fluid Motion. The Parabolic Press, Stanford California (1982)
28. Roberts, B.M.: Jean Claude Eugéne Péclet. http://www.hevac-heritage.org/built_environment/biographies/surnames_M-R/peclet/P1-PECLET.pdf
29. Billington, N.S., Roberts, B.M.: Building Services Engineering: A Review of Development. Pergamon Press, Oxford (1982)
30. Zhao, C., Hobbs, B.E., Ord, A.: Convective and Advective Heat Transfer in Geological Systems. Springer, Berlin (2008)
31. Whipp, D.: Thermal processes in the lithosphere. https://matskut.helsinki.fi/bitstram/handle/123456789/831/Lecture%208.5%20-%20The%20Peclet%20number.pdf?sequence=1
32. Eckert, E.R.G.: Wärme- und Stoffaustausch. Springer, Berlin (1966)
33. Minkowycz, W.J.: Professor Ernst R.G. Eckert (1904-2004). Int. J. Heat Mass Trans. Elsevier **49** (2006)
34. Pfender, E.: Memorial Tributes. The National Academic Press, vol. 11, 108–113 (2007)
35. Inger, G.R.: Scaling Nonequilibrium-Reacting Flows: The Legacy of Gerhard Damköhler. J. Spacecraft Rockets **38**(2) (2001)
36. Wicke, E.: Gerhard Damköhler - Begründer der Chemischen Reaktionstechnik. Chem.-Ing.-Tech. **56**(12) (1984)
37. Weiland, C.: Computational Space Flight Mechanics. Springer, Berlin (2010)
38. Weiland, C.: Aerodynamic Data of Space Vehicles. Springer, Berlin (2014)
39. Hirschel, E.H., Weiland, C.: Selected Aerothermodynamic Design Problems of Hypersonic Flight Vehicles. Springer, Berlin and AIAA (Progress in Astronautics and Aeronautics), Reston USA (2009)

40. Hagemann, G., Knab, O.: Private communications. EADS - Astrium Space Transportation (2010)
41. Abgrall, R., Désidéri, J.-A., Glowinski, R., Mallet, M., Périaux, J. (eds.): Hypersonic Flows for Reentry Problems, vol. III. Springer, Berlin (1992)
42. Wegener, P.P., Buzyna, G.: J. Fluid Mech. **37**, 325–335 (1969)
43. Schmidt, E. : Ernst Kraft Wilhem Nusselt (1882 bis 1957). Forschungen für die Wärmetechnik. In: Buchheim, G., Sonnemann, R., (eds.) Lebensbilder von Ingenieurwissenschaftlern. Leibzig (1989)
44. Nusselt, W.: Technische Thermodynamik. de Gruyter Verlag, Berlin (1934)
45. VDI Wärmeatlas: VDI Gesellschaft Verfahrenstechnik und Chemieingenieurswesen, (Eds.). Springer-Vieweg, Berlin (2013)
46. Désidéri, J.-A., Glowinski, R., Periaux, J. (eds.): Hypersonic Flows for Reentry Problems, vols. I and II. Springer, Berlin (1991)
47. Schroeder, W., Menne, S.: Hypersonic Delta-Wing Flow, Case VII.4. In: Abgrall, R., Désidéri, J.-A., Glowinski, R., Mallet, M., Périaux, J. (eds.) Hypersonic Flows for Reentry Problems, vol. III. Springer, Berlin (1992)
48. Delery, J., Coet, M.-C.: Experiments on Shock/Boundary-Layer Interactions Producd by Two-Dimensional Ramps and Three-Dimensional Obstacles. In: Désidéri, J.-A., Glowinski, R., Périaux, J. (eds.) Hypersonic Flows for Reentry Problems, vol. II. Springer, Berlin (1991)
49. Hottel, H.C.: Bibliographical memoirs of Warren Kendall Lewis. In: National Academy of Sciences, National Academic Press, Washington D.C., (1996)
50. Lewis, W.K.: J. Indust. Eng. Chem. **1**, 522–533 (1909)
51. Vad, J., Lajos, T., Schilling, R. (eds.): Modelling Fluid Flow. Springer, Berlin (2004)
52. Kurzweil, P.: Das Vieweg Einheiten Lexikon. Vieweg-Verlag, Braunschweig (1999)
53. https://authors.libraray.caltech.edu/61287/3mmc2.pdf
54. http://www.thermopedia.com/content/1109
55. https://tu-dresden.de/ing/maschinenwesen/ilr/ressourcen/dateien/tfd/studium/dateien/Thermofluiddynamik_F.pdf?lang=de
56. Schütz, Th. (ed.): Hucho - Aerodynamic des Automobils. Springer-Vieweg, Braunschweig (2013)
57. Hitzel, S.: Private communication. EADS Military Aircraft Systems, Ottobrunn (2002)
58. Prandtl, L.: Über Flüssigkeitsbewegung bei sehr kleiner Reibung. Verh. d. III. Intern. Math. Kongress, Heidelberg, Germany (1904). (B.G. Teubner Verlag, Leibzig, 485–491 (1905))
59. Meier, G.E.A. (ed.): Ludwig Prandtl, ein Führer in der Strömungslehre. Vieweg Verlag, Braunschweig (2000)
60. Rotta, J.C.: Die Aerodynamische Versuchsanstalt in Göttingen, ein Werk Ludwig Prandtls. Verlag Vandenhoeck & Ruprecht, Göttingen (1990)
61. Christen, D.S.: Praxiswissen der chemischen Verfahrenstechnik. Springer, Berlin (2010)
62. Herwig, H.: Wärmeübertragung A-Z. VDI Buch. Springer, Berlin (2000)
63. Schlichting, H., Gersten, K.: Grenzschichttheorie, 10th edn. Springer, Berlin (2006)
64. Jischa, M.: Konvektiver Impuls-, Wärme- und Stoffaustausch. Springer Fachwissen, Berlin (1982)
65. Herival, J.: Joseph Fourier: the man and the physicist. Clarendon Press, Oxford (1975)
66. Nowlan, R.: A chronicle of mathematical people. pdf - file
67. Fourier, J.-B.-J.: Théorie analytique de la chaleur. Firmin Didot et Fils (1822)
68. Struik, D.J.: Joseph Fourier, French Mathematician. Encyclopeadia Britannica
69. Böckh v., P.: Wärmeübertragung Grundlagen und Praxis. Springer, Berlin (2006)
70. N.N.: Unsteady State Heat Transfer. http://www3.nd.edu/~amoukasi/CBE358_Lab1/HT3_Unsteady%20State%20Heat%20Transfer_f.pptx
71. Kassab, A.: Application of Series in Heat Transfer - transient heat conduction (2008). https://excel.ucf.edu/classes/2008/Spring/appsII/chapter_7b_Sp_08.pdf
72. Strouhal, V.: Über eine besondere Art der Tonerregung. Annalen der Physik und Chemie, vol. 5, Leipzig, Germany. 216–251 (1878)

73. Költzsch, P.: Fragmente aus der Geschichte der Strömungsakustik. University of Dresden, Germany. https://pub.dega-akustik.de/DAGA_1999-2008/data/articles/00121.pdf
74. Corda, S.: Introduction to Aerospace Engineering with Flight Test Perspective. Wiley, New York (2017)
75. Würz, W., Gaisbauer, U.: Strouhal Zahl, University of Stuttgart, (2014). https://www.iag-stuttgart.de/abteilungen/laminarwindkanal/pdf_laminar/Umdrucke_Messtechnik/06_Aehnlichkeitszahlen_Messtechnik_WS2014.pdf
76. Billah, K.Y., Scanlan, R.H.: Resonance, Tacoma narrows bridge failure, and undergraduate physics textbooks. Am. J. Phys. **59** (1991)
77. https://en.wikipedia.org/wiki/Tacoma_Narrows_Bridge_(1940)
78. Nobel Lectures, Physics 1901–1921. Elsevier Publishing Company, Amsterdam (1967). https://www.nobelprize.org/nobel_prizes/physics/laureates/1904/
79. O'Connor, J.J., Robertson, E.F.: John William Strutt (Lord Rayleigh). http://www-groups.dcs.st-and.ac.uk/history/Printonly/Rayleigh.html
80. Lindsay, R.B.: John William Strutt, 3rd Lord Rayleigh. https://www.britannica.com/print/article/492464
81. Görtler, H.: Dimensionsanalyse. Theorie der physikalischen Dimensionen und Anwendungen. Springer, Berlin (1975)
82. Ahlers, G., Grossmann, S., Lohse, D.: Hochpräzision im Kochtopf. Phys. J., vol. 2, Wiley-VCH Verlag (2002)
83. Bénard, H.: Les tourbillions cellulaires dans une nappe liquide. Revue Générale des Science, vol. 11, 1261–1271 and 1309–1328 (1900)
84. Strutt, J.W.: On the convection currents in a horizontal layer of fluid when the higher temperature is on the under side. Phil. Mag. **2**, 833–844 (1916)
85. Merker, G.P.: Konvektive Wärmeübertragung. Springer, Berlin (1987)
86. Voit, C.: Nekrolog auf Sir George Gabriel Stokes. Sitzungsberichte der mathematisch-physikalischen Klasse der K.B. Akademie der Wissenschaften, München, Band XXXIII (1903)
87. Mathematical and Physical Papers by George Gabriel Stokes. https://www.cambridge.org/core/books/mathematical-and-physical-papers/58B93740612EFCB1C365A1DCC3
88. Krause, E.: Strömungslehre, Gasdynamik und aerodynamisches Laboratorium. Teubner Verlag, Stuttgart (2003)
89. Stokes, G.G.: Trans. Camb. Phil. Soc. **2** (Part II), 8–106 (1851)
90. https://www.math.nyu.edu/faculty/childres/chpseven.PDF
91. Cromer, D., Pruisner, L.: Snow, Rain and the Stokes number. s3.danielcromer.com/resources/SnowRainStokesNo.pdf
92. http://www.diss.fu-berlin.de/diss/servlet/MCRFileNodeServlet/FUDISS_derivate_000000003104/02_Kap_p9-p52.pdf
93. Dillmann, A.: Future Perspectives of Experimental Aerodynamics. DLR Göttingen. Paper presented at the celebration of E.H. Hirschel's 80th birthday, DLR Cologne (2014)
94. Adam, H., Steckelmacher, W.: Vacuum Science and Technology, -Pioneers of the 20th Century-. Redhead, P.A. (ed.). AIP Press, Maryland (1994)
95. von Kármán, Th.: ZAMM **3**, 395–396 (1923)
96. Hänel, D.: Molekulare Gasdynamik. Springer, Berlin (2004)
97. Stemmer, C.: Hyperschallströmungen. Vorlesung TU München, Germany. https://www.aer.mw.tum.de/fileadmin/tumwaer/www/pdf/lehre/hyperschallstroem/skript.pdf
98. Frankovic, B., Pohl, G.: Peter Salcher and Ernst Mach. Proc. Int. Symp, Rijeka, Croatia (2004)
99. http://www.scientificlib.com/en/Physics/Biographies/ErnstMach.html
100. Ackeret, J.: Habilitationsschrift, ETH Zürich (1928)
101. Mach, E., Salcher, P.: Photographische Fixierung der durch Projectile in der Luft eingeleiteten Vorgänge. Sitzungsberichte der Kaiserlichen Akademie der Wissenschaften, Part 2 **95**, 764–781 (1887)
102. Mach, E.: Über die Abbildung der von Projectilen mitgeführten Luftmasse durch Momentphotographie. Anzeiger der Kaiserlichen Akademie der Wissenschaften **23**, 136 (1886)

103. Mach, E., Salcher, P.: Optische Untersuchung der Luftstrahlen. Sitzungsberichte der Akademie der Wissenschaften, XCVIII Band, Wien (1889)
104. Gross, A., Weiland, C.: Numerical Simulation of Separated Cold Gas Nozzle Flows. J. Propul. Power **20**(3), 509–519 (2004)
105. O'Connor, J.J., Robertson, E.F.: Josef Stefan. http://www-history.mcs.st-andrews.ac.uk/Biographies/Stefan_Josef.html
106. Jiji, L.M.: Heat Conduction. Springer, Berlin (2009)
107. Liu, J: Stability of viscoplastic flow. https://www.whoi.edu/fileserver.do?id=283458&pt=10&p=17275
108. Osswald, T.A.: Understanding Polymer Processing. Hanser Verlag, Munich (2011)
109. Baehr, H.D., Stephan, K.: Heat and Mass Transfer. Springer, Berlin (2006)
110. Kunes, J.: Dimensionless Physical Quantities in Science and Engineering. Elsevier Publishing Company, London (2012)
111. Grachev, A.A., Andreas, E.L., Fairall, Ch.W., Guest, P.S., Persson, P.O.G.: The Critical Richardson Number and Limits of Applicability of Local Similarity Theory in the Stable Boundary Layer. https://arvix.org/ftp/arvix/papers/1202/1202.5066.pdf
112. Roshko, A.: Experiments on the flow past a circular cylinder at very high Reynolds number (1960). https://pdfs.semanticscholar.org/2b0e/17ef8b142cd1100ec608956c75aa5ce24c.pdf
113. Rattner, A., Bohren, J.: Heat and Mass Correlations. https://www.stwing.upenn.edu/~salexa/Documents/Correlations.pdf

Chapter 7
Gasdynamic Similarity

As we have already discussed in Chaps. 2, 3 and 4 the calculus and the aim of the procedure and the determination of dimensionless numbers is to reduce the amount of variables, which describe fluid dynamical and thermal processes. As long as the scientists and engineers were not able to solve satisfactorily the governing equations for fluid dynamic/thermodynamic fields this approach was one way to learn more about the peculiarities of all kind of flow and heat transfer problems. Further this people was engaged up to the 60s of the last century—and in particular between the two World Wars—with the challenge to find theoretical ways to make analytical solutions of the governing equations of the fluid dynamics possible. These equations are the full potential equation, the Euler equation or—on the highest level—the Navier–Stokes equations. But at that time analytical solutions were only possible, if very far-reaching restrictions and simplifications were introduced. This led to reduced forms of the potential equation, which was, for example, linearized by series expansions for subsonic and supersonic flows past thin or slender bodies (e.g.: profiles and wings) at small angles of attack. For other Mach number regimes linearization was not possible (transonic) or was not leading to the desired result (hypersonic) as is discussed below.

7.1 The Goal of the Gasdynamic Similarity

The objective of the derivation of similarity rules was to generate for a family of body contours a reference solution either experimentally or analytically, for example by applying the method of singularities for incompressible flow[1] ($M_\infty = 0.0$) and

[1] In compressible, supersonic flow the solution of d'Alembert (hyperbolic partial differential equation) is used, [1].

C. Weiland, *Mechanics of Flow Similarities*,
https://doi.org/10.1007/978-3-030-42930-0_7

to transfer these solutions to higher subsonic Mach numbers, and the same holds for the other Mach number regimes.

Today, this aspect lost importance, since modern numerical methods are able to solve the Euler equations or even the Navier–Stokes equations in three dimensions with often a satisfactory degree of accuracy. Further there are exciting advances in modern experimental techniques, like new wind tunnels, where increased Reynolds numbers and a wide range of Mach numbers can be realized. These facilities are equipped with modern measurement systems, like **P**article **I**nduced **V**elocimetry (PIV) and **P**ressure **S**ensitive **P**aint (PSP) systems as well as infrared and liquid crystal thermographic methods.

Nevertheless, below approximate theories are summarized, which were developed for subsonic, transonic, supersonic and hypersonic Mach numbers with the goal to formulate similarity rules and approximate relations for the determination of flow field quantities like the pressure coefficient, the lift coefficient etc..

Approximate theories including similarity rules were developed for

(A) Two- and three dimensional flows for subsonic and supersonic Mach numbers past slender body contours and small angles of attack using the linearized small perturbation potential equation. Essentially the scientific work was performed by: L. Prandtl, H. Glauert, B. Göthert, J. Ackeret, A. Busemann [2–6].

Ludwig Prandtl
1875 - 1953

Hermann Glauert
1892 - 1934

Jakob Ackeret
1898 - 1981

Bernhard Goethert
1907 - 1988

Adolf Busemann
1901 - 1986

(B) Axisymmetric flows past slender bodies of revolution without angle of attack using the linearized small perturbation potential equation in cylindrical coordinates. The scientific work essentially was made by: K. Oswatitsch [7].

Klaus Oswatitsch
1910 - 1993

(C) Two- and three dimensional transonic flows ($1-\varepsilon \leqq M_\infty \leqq 1+\varepsilon$, $\varepsilon = \mathcal{O}(0.1)$) past slender bodies with small angles of attack using the non-linear small perturbation potential equation. The scientific work essentially was performed by: Th. von Kármán, K. Oswatitsch, G. Guderley [7–9].

Theodore von Kármán
1881 - 1963

Gottfried Guderley
1910 - 1997

(D) Two- and three dimensional hypersonic flows past slender bodies and/or small angles of attack as well as blunt bodies. All that on the basis of linearized and non-linearized approximate equations. The scientific work essentially was made by: H.S. Tsien, K. Oswatitsch, J. Zierep [10]–[12].

H.S. Tsien
1911 - 2009

Jürgen Zierep
1929 –

In Chap. 1 we already had referred to, that the application of most of the similarity laws (or dimensionless numbers) requires geometrical similarity of the bodies, which are passed by the fluid (aircrafts, spacecrafts, etc.) or the devices, which are flowed through (inlets, hydraulic engines, etc.), Chap. 8. Full geometrical similarity means, see Sect. 8.1

$$\mathbf{r}_1 = \alpha_1 \mathbf{r}_2 \quad \text{or} \quad x_1 = \alpha_1 x_2,\ y_1 = \alpha_1 y_2,\ z_1 = \alpha_1 z_2\,.$$

The consideration of the gasdynamic similarity does not require the full geometrical similarity, but uses affine transformations of the geometry of the bodies, which means that the bodies will be warped. The coordinates describing the surface of such a body transform then by

$$x_1 = x_2,\ y_1 = \beta_1 y_2,\ z_1 = \beta_2 z_2 \qquad \beta_1, \beta_2 > 0\,.$$

7.2 Derivation of the Small Perturbation Equation

The basis of the consideration of gasdynamic similarity for subsonic and supersonic flows is the linearized potential equation. For completeness we demonstrate the derivation of this equation by starting from the full Navier–Stokes equations, Eqs. (4.5), (4.7) and (4.11), Sect. 4.2, using the appropriate assumptions. Firstly, we drop all the dissipative and diffusive terms (including the heat sources) from the Navier–Stokes equations and obtain the Euler equations describing inviscid flow fields.

$$
\begin{aligned}
\text{continuity}\ & \frac{\partial \rho}{\partial t} + \frac{\partial \rho u}{\partial x} + \frac{\partial \rho v}{\partial y} + \frac{\partial \rho w}{\partial z} = 0\,,\\
\text{x - momentum}\ & \frac{\partial}{\partial t}(\rho u) + \frac{\partial}{\partial x}(\rho u^2 + p) + \frac{\partial}{\partial y}(\rho u v) + \frac{\partial}{\partial z}(\rho u w) = 0\,,\\
\text{y - momentum}\ & \frac{\partial}{\partial t}(\rho v) + \frac{\partial}{\partial x}(\rho v u) + \frac{\partial}{\partial y}(\rho v^2 + p) + \frac{\partial}{\partial z}(\rho v w) = 0\,,\\
\text{z - momentum}\ & \frac{\partial}{\partial t}(\rho w) + \frac{\partial}{\partial x}(\rho w u) + \frac{\partial}{\partial y}(\rho w v) + \frac{\partial}{\partial z}(\rho w^2 + p) = 0\,,\\
\text{energy}\ & \frac{\partial}{\partial t}(\rho e) + \frac{\partial}{\partial x}\left((\rho e + p)u\right) + \frac{\partial}{\partial y}\left((\rho e + p)v\right) +\\
& \frac{\partial}{\partial z}\left((\rho e + p)w\right) = 0\,,
\end{aligned}
\tag{7.1}
$$

The energy equation can be reformulated to

$$
\frac{ds}{dt} = \frac{\partial s}{\partial t} + \mathbf{v} \cdot grad\ s = 0\,,
$$

with s being the entropy, which is a remarkable result. Secondly, we assume that the flow field contains no vorticity

$$
rot\ \mathbf{v} = 0\,.
$$

Then a potential function Φ exists with

$$
\mathbf{v} = grad\ \Phi \quad \text{and} \quad rot\ grad\ \Phi = 0\,.
$$

In an irrotational flow field (vorticity free) the momentum equations and the energy equation are no longer independent of each other. Further the entropy s is a constant in the whole flow field. Therefore only the continuity equation remains as independent differential equation and we have

$$
\frac{\partial \rho}{\partial t} + \frac{\partial \rho u}{\partial x} + \frac{\partial \rho v}{\partial y} + \frac{\partial \rho w}{\partial z} = 0 \tag{7.2}
$$

or

$$
\frac{\partial \rho}{\partial t} + \frac{\partial \rho \Phi_x}{\partial x} + \frac{\partial \rho \Phi_y}{\partial y} + \frac{\partial \rho \Phi_z}{\partial z} = 0\,.
$$

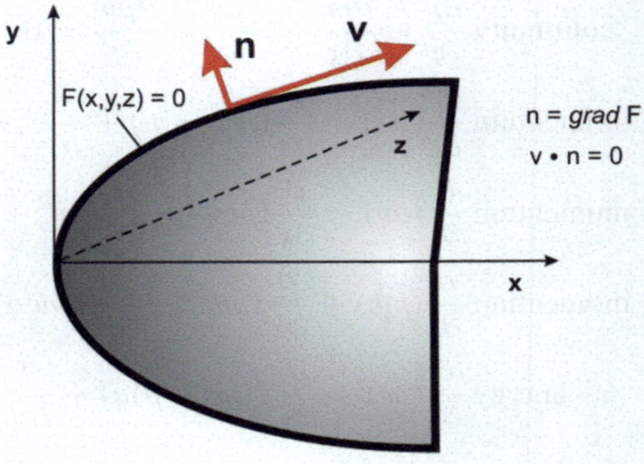

Fig. 7.1 Boundary condition for inviscid flows using the function of the surface contour $F(x, y, z) = 0$

This is the conservatively formulated potential equation. Together with the algebraic equations[2] for the pressure p and the density ρ

$$\rho = \left(1 - \frac{\gamma - 1}{2} M_\infty^2 (u^2 + v^2 + w^2 - 1)\right)^{\frac{1}{\gamma-1}} , \tag{7.3}$$
$$p = \frac{\rho^\gamma}{\gamma M_\infty^2} ,$$

the five unknowns describing an inviscid flow field are identified. The boundary condition for inviscid flow consists in the fact that the flow velocity vector has to be tangential to the body surface (no inflow into the body), Fig. 7.1. This is valid for the Euler equations and the full potential equation as well:

$$\mathbf{v} \cdot \mathbf{n} = 0, \quad \text{with} \quad \mathbf{n} = \frac{grad\ F}{|grad\ F|} \quad \text{and} \quad F(x, y, z) = 0 , \tag{7.4}$$

where **n** is the body normal and $F(x, y, z)$ describes the body contour.

Remark Often the potential equation is used in a non-conservative formulation. For a steady flow field we present the derivation. Starting with Eq. (7.2), which has to be evaluated, and using Eq. (7.3) for the density ρ as well as $rot\ \mathbf{v} = 0$ and $\mathbf{v} = grad\ \Phi$ yields

---

[2]These equations are derived from the energy equation of the flow filament theory $\frac{\gamma}{\gamma-1}\frac{p_\infty}{\rho_\infty} + \frac{v_\infty^2}{2} = \frac{\gamma}{\gamma-1}\frac{p}{\rho} + \frac{v^2}{2}$ and the isentropic relation $\frac{p}{p_\infty} = \left(\frac{\rho}{\rho_\infty}\right)^\gamma$.

$$\left(1-\frac{u^2}{a^2}\right)\Phi_{xx}+\left(1-\frac{v^2}{a^2}\right)\Phi_{yy}+\left(1-\frac{w^2}{a^2}\right)\Phi_{zz}- \tag{7.5}$$
$$-2\frac{uv}{a^2}\Phi_{xy}-2\frac{uw}{a^2}\Phi_{xz}-2\frac{vw}{a^2}\Phi_{yz}=0\,,$$

with a being the speed of sound. This is the non-conservatively formulated full potential equation describing steady, three dimensional flows.

Finally, we consider small perturbations of the velocity with $u = u_\infty + \delta u$ and $v = v_\infty + \delta v \ll u_\infty$, $w \ll u_\infty$, $\delta u \ll u_\infty$ as well as $u \approx u_\infty$, $a = a_\infty$, which is inserted into Eq. (7.5), and take into account only first order terms. Then we obtain the linearized potential equation, valid for flow fields with small configurational variations (e.g.: thin wings or profiles) and small angles of attack:

$$(1-M_\infty^2)\varphi_{xx}+\varphi_{yy}+\varphi_{zz}=0\,, \tag{7.6}$$

with φ being the perturbation potential.[3] In subsonic flow this equation is of elliptic type and in supersonic flow of hyperbolic type, which requires fundamentally different solution procedures, [1, 9, 13].

The two dimensional version of Eq. (7.6), when transformed by $x = \bar{x}$ and $\beta y = \bar{y}$, $\beta = \sqrt{1 - M_\infty^2}$, the Prandtl–Glauert factor, yields the classical Laplace equation:

$$\varphi_{\bar{x}\bar{x}}+\varphi_{\bar{y}\bar{y}}=0\,. \tag{7.7}$$

This equation is typically solved in fluid mechanics by the method of singularities, [1, 9]. Further for incompressible flows ($M_\infty = 0$), Eq. (7.6) is also of Laplace type. Often the incompressible flow serves in the frame of the gasdynamic similarity theory as reference solution.

The three dimensional extension of the method of singularities is the Panel method, which was used in the past for design purposes in the aircraft industry.[4]

For the two dimensional case the boundary condition reads with the contour function $F(x, y) = f(x) - y = 0$, $\quad y = f(x)$ and $u\partial F/\partial x + v\partial F/\partial y = 0$ (Eq. (7.4)):

$$\frac{dF(x,y)}{dx}=\frac{v}{u}\quad\text{and for small perturbations}\quad\Longrightarrow\frac{\varphi_y}{u_\infty+\varphi_x}\,, \tag{7.8}$$

with φ being the potential of the perturbations.

The typical strategy for solving the Eq. (7.7) for an unsymmetrical airfoil with the boundary condition Eq. (7.8) by the singularity method consists in splitting up the

[3]The small perturbation potential equation for unsteady flows reads
$(1 - M_\infty^2)\varphi_{xx} + \varphi_{yy} + \varphi_{zz} = 1/a_\infty^2(\varphi_{tt} + 2\varphi_x\,\varphi_{xt})$.

[4]Due to the drawbacks of this method and the advent of very powerful computers as well as advanced numerical algorithms and grid generation procedures, today more sophisticated methods like Euler codes or even Navier–Stokes codes are employed.

problem into a thickness-, an angle of attack- and a camber part, where the splitted boundary conditions[5] are

$$\left.\frac{dF(x,y)}{dx}\right|_{thickness} = \frac{\hat{\varphi}_y}{u_\infty + \hat{\varphi}_x}\,, \qquad \left.\frac{dF(x,y)}{dx}\right|_{angle\ of\ attack} = \frac{v_\infty + \check{\varphi}_y}{\check{\varphi}_x}\,.$$

For the camber effect a linear combination of $dF(x, y)/dx$ for the upper and the lower sides of the airfoil is employed. The analytical integration along the profile surface carried out for each splitted system, using the appropriate singularities, results in the related u and v distributions, which are superimposed due to the linearity of the equation in order to obtain the global solution. For details see e.g. [9].

For the formulation of the similarity rule we choose the specific contour function:

$$F(x,y) = f(x) - \frac{l}{d}y = 0\,, \tag{7.9}$$

with l being a reference length and d the maximal thickness of a thin body (e.g.: a wing profile).

The boundary condition reads then using Eq. (7.8):

$$\frac{\partial\varphi}{\partial y} = u\,\frac{d}{l}\,\frac{df(x)}{dx} = u\,\frac{dy}{dx}\,. \tag{7.10}$$

7.3 Similarity Rules for Two and Three Dimensional Flows at Subsonic and Supersonic Mach Numbers (A)

We now consider a first flow, for which the perturbation potential equation, the boundary condition and the pressure coefficient[6] are defined by, [9–16]

$$(1 - M_{\infty 1}^2)\,\varphi_{1,x_1x_1} + \varphi_{1,y_1y_1} = 0\,, \tag{7.11}$$

$$\frac{\partial\varphi_1}{\partial y_1} = u_1\frac{d_1}{l_1}\,\frac{df(x_1)}{dx_1}\,, \tag{7.12}$$

$$c_{p1} = -\frac{2}{u_1}\,\frac{\partial\varphi_1}{\partial x_1}\,. \tag{7.13}$$

For a second flow it holds

---

[5]The perturbation potential is also splitted in a thickness part $\hat{\varphi}$ and an angle of attack part $\check{\varphi}$.

[6]The pressure coefficient (Eq. (7.13)) was derived from its definition $c_p = \dfrac{p - p_\infty}{1/2\rho_\infty u_\infty^2}$ for flows with small perturbations [1, 9].

$$(1 - M_{\infty 2}^2)\, \varphi_{2,x_2x_2} + \varphi_{2,y_2y_2} = 0\,, \tag{7.14}$$

$$\frac{\partial \varphi_2}{\partial y_2} = u_2 \frac{d_2}{l_2} \frac{df(x_2)}{dx_2}\,, \tag{7.15}$$

$$c_{p2} = -\frac{2}{u_2} \frac{\partial \varphi_2}{\partial x_2}\,. \tag{7.16}$$

Applying the transformation between x_1, y_1, φ_1 and x_2, y_2, φ_2

$$x_2 = x_1\,, \qquad y_2 = a y_1\,, \qquad \varphi_1(x_1, y_1) = C\, \varphi_2(x_2, y_2)\,, \tag{7.17}$$

we obtain with Eq. (7.11)

$$(1 - M_{\infty 1}^2)\, \varphi_{2,x_2x_2} + a^2\, \varphi_{2,y_2y_2} = 0\,. \tag{7.18}$$

This agrees only with Eq. (7.14) if

$$a^2 = \frac{1 - M_{\infty 1}^2}{1 - M_{\infty 2}^2}$$

holds. From the boundary conditions Eqs. (7.12 and 7.15) one finds

$$\begin{aligned} \frac{d\,f(x_1)}{dx_1} &= \frac{l_1}{d_1 u_1} \frac{\partial \varphi_1}{\partial y_1} = \frac{l_1 a}{d_1 u_1}\, C\, \frac{\partial \varphi_2}{\partial y_2} = \\ \frac{d\,f(x_2)}{dx_2} &= \frac{l_2}{d_2 u_2} \frac{\partial \varphi_2}{\partial y_2}\,, \end{aligned} \tag{7.19}$$

because the transformation above does not change the derivative $df(x_i)/dx_i$, and therefore it follows for C

$$C = \frac{1}{a} \frac{d_1}{l_1} \frac{l_2}{d_2} \frac{u_1}{u_2}\,. \tag{7.20}$$

For the pressure coefficients one obtains

$$\begin{aligned} c_{p1} &= \frac{1}{a} \frac{d_1}{l_1} \frac{l_2}{d_2}\, c_{p2}\,, \\ c_{p1} \frac{l_1}{d_1} \sqrt{1 - M_{\infty 1}^2} &= c_{p2} \frac{l_2}{d_2} \sqrt{1 - M_{\infty 2}^2}\,. \end{aligned} \tag{7.21}$$

Note For the reference flow case 1 often the Mach number $M_{\infty 1} = 0$ (incompressible flow, $\sqrt{1 - M_{\infty 1}^2} = 1$) for the subsonic flow regime is used and $M_{\infty_1} = \sqrt{2}$ for the supersonic flow regime $\left(\sqrt{|1 - M_{\infty 1}^2|} = 1\right)$.

Prandtl–Glauert Rule

If the thickness ratio of the two bodies is identical ($d_1/l_1 = d_2/l_2$), then the pressure coefficient c_{p2} increases with increasing $1/\sqrt{1 - M_{\infty 2}^2}$, [3]

$$c_{p2} = c_{p1} \frac{\sqrt{1 - M_{\infty 1}^2}}{\sqrt{1 - M_{\infty 2}^2}} . \tag{7.22}$$

Example Assumed the pressure coefficient c_{p1} for the first flow with $M_{\infty 1} = 0.35$ is known, then the pressure coefficient c_{p2} for the second flow with $M_{\infty 2} = 0.75$ around the constant body ($d_2/l_2 = d_1/l_1$) amounts to $c_{p2} = 1.416\, c_{p1}$.

Göthert Rule

The pressure coefficient c_{p2} increases about $1/(1 - M_{\infty 2}^2)$, if the thickness ratio $\frac{d_2}{l_2}/\frac{d_1}{l_1}$ grows with $1/\sqrt{1 - M_{\infty 2}^2}$. This behaviour is also called the analogy of streamlines, [4]:

$$c_{p2} = c_{p1} \frac{\sqrt{1 - M_{\infty 1}^2}}{1 - M_{\infty 2}^2} . \tag{7.23}$$

Example Assumed the pressure coefficient c_{p1} from the first flow with $M_{\infty 1} = 0.35$ is known, then the pressure coefficient c_{p2} for the second flow with $M_{\infty 2} = 0.75$ around the body is increased by $\frac{d_2}{l_2} = \frac{d_1}{l_1} \frac{1}{\sqrt{1 - M_{\infty 2}^2}}$, Fig. 7.2a, amounts to $c_{p2} = 2.141\, c_{p1}$.

Analogy of Potential Lines

We consider the case that in both flows the pressure coefficients are not changed ($c_{p2} = c_{p1}$). This means that $\frac{l}{d}\sqrt{1 - M_\infty^2}$ is a constant (Eq. (7.21)) and further the potentials are identical, because the perturbations of the velocity components are identical ($\varphi_2 = \varphi_1$, Eq. (7.17)).

Example For our two flows with $M_{\infty 1} = 0.35$ and $M_{\infty 2} = 0.75$ the correlation $\frac{\sqrt{1 - M_{\infty 1}^2}}{\sqrt{1 - M_{\infty 2}^2}}$ amounts to 1.416. Since $\frac{l}{d}\sqrt{1 - M_\infty^2}$ has to be constant the relation $\frac{d_2}{l_2} = \frac{1}{1.416} \frac{d_1}{l_1}$ holds, Fig. 7.2b.

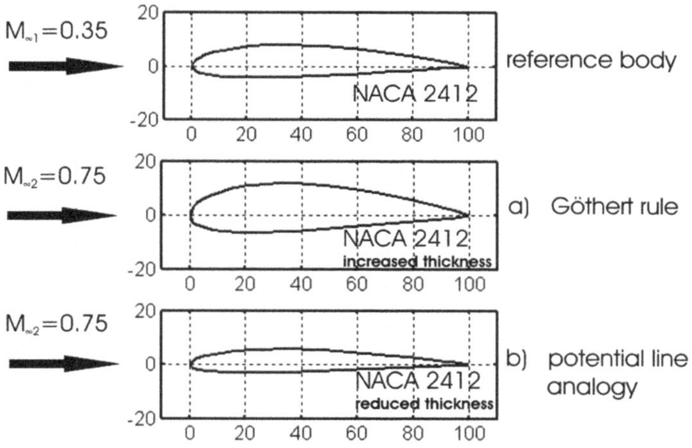

Fig. 7.2 NACA 2412 profile: a) the Göthert rule increases the thickness, if $l_2 = l_1$, b) the potential line analogy reduces the thickness, if $l_2 = l_1$

Three Dimensional Flows

The three dimensional flow case can be considered in the same way as the two dimensional one. With Eq. (7.6) and the transformation of the coordinates (see Eq. (7.17))

$$x_2 = x_1\,, \quad y_2 = a y_1\,, \quad z_2 = a z_1\,, \quad \varphi_1(x_1, y_1, z_1) = C\,\varphi_2(x_2, y_2, z_2)\,, \tag{7.24}$$

we find for the transformed coordinates

$$\begin{aligned} x_2 &= x_1\,, \\ y_2\sqrt{|1 - M_{\infty 2}^2|} &= y_1\sqrt{|1 - M_{\infty 1}^2|}\,, \\ z_2\sqrt{|1 - M_{\infty 2}^2|} &= z_1\sqrt{|1 - M_{\infty 1}^2|}\,, \end{aligned} \tag{7.25}$$

and with $M_{\infty 1} = 0, \quad 0 < M_{\infty 2} < 1.0$ we obtain $y_2 = y_1 \dfrac{1}{\sqrt{1 - M_{\infty 2}^2}}$, which leads to an enlargement of the aspect ratio, Fig. 7.3. In supersonic flow with $M_{\infty 1} = \sqrt{2}$, $M_{\infty 2} > \sqrt{2}$ we have $y_2 = y_1 \dfrac{1}{\sqrt{|1 - M_{\infty 2}^2|}}$, which reduces the aspect ratio, Fig. 7.4.

Another solution is given with $x_2 = x_1/a\,, \quad y_2 = y_1\,, \quad z_2 = z_1$, where in subsonic flow the x coordinate is reduced by $x_2 = x_1\sqrt{1 - M_{\infty 2}^2}$ at constant y coordinate $y_2 = y_1$ (and also $z_2 = z_1$), Fig. 7.5.

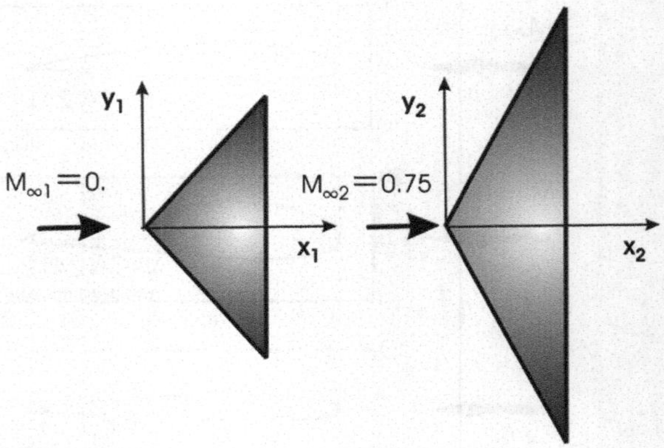

Fig. 7.3 Flow similarity: change of the aspect ratio due to y variation ($x_2 = x_1$) in subsonic regime for $M_{\infty 1} = 0. \longrightarrow M_{\infty 2} = 0.75$

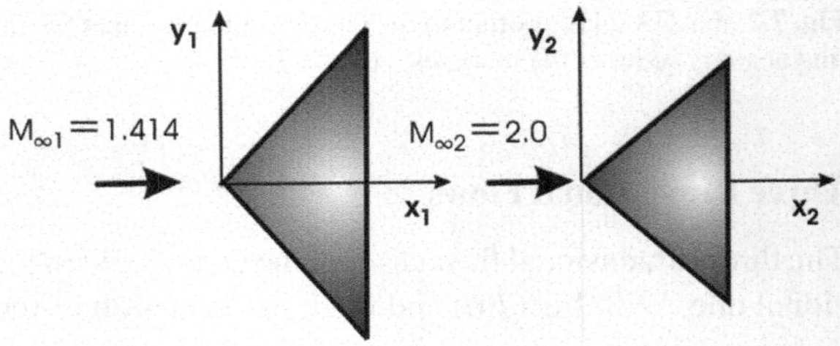

Fig. 7.4 Flow similarity: change of the aspect ratio due to y variation ($x_2 = x_1$) in supersonic regime for $M_{\infty 1} = \sqrt{2} \longrightarrow M_{\infty 2} = 2.0$

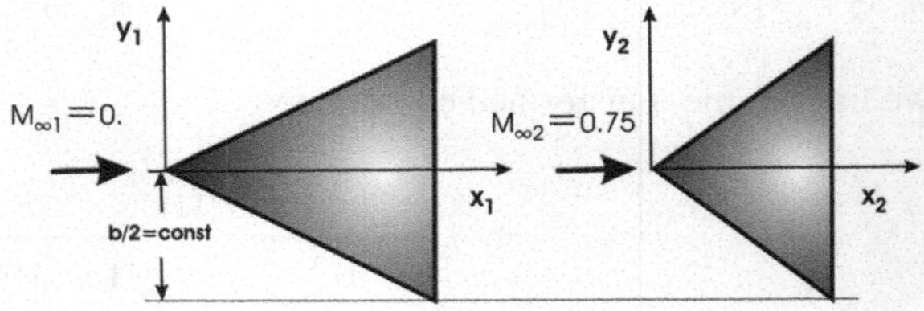

Fig. 7.5 Flow similarity: change of the aspect ratio due to x variation ($y_2 = y_1$) in subsonic regime for $M_{\infty 1} = 0. \longrightarrow M_{\infty 2} = 0.75$

7.4 Similarity Rules for Axisymmetric Flows Past Slender Bodies of Revolution (B)

When a gas streams around a body of revolution without angle of attack the flow is axisymmetric. For a thin body the three dimensional linearized potential equation, Eq. (7.6), formulated in Cartesian coordinates x, y, z can be employed, which is then transformed to cylindrical coordinates x, r, ϑ

$$(1 - M_\infty^2)\frac{\partial^2 \varphi}{\partial x^2} + \frac{1}{r}\frac{\partial}{\partial r}\left(r\frac{\partial \varphi}{\partial r}\right) + \frac{1}{r^2}\frac{\partial^2 \varphi}{\partial \vartheta^2} = 0 \,. \qquad (7.26)$$

Since the flow past a body of revolution with zero angle of attack ($\alpha = 0°$) in the cylindrical coordinate system behaves two dimensionally (axisymmetric) Eq. (7.26) reduces to

$$(1 - M_\infty^2)\frac{\partial^2 \varphi}{\partial x^2} + \frac{\partial^2 \varphi}{\partial r^2} + \frac{1}{r}\frac{\partial \varphi}{\partial r} = 0\,. \tag{7.27}$$

The boundary condition Eq. (7.8), when applied to cylindrical coordinates fails for $r \to 0$ (in the vicinity of the x axis) due to the existence of singularities of the velocity components. To overcome this problem the relation

$$r\frac{dF}{dx} = \frac{rv}{u}\,, \tag{7.28}$$

is used, where rv in the vicinity of $r \to 0$ is only a function of the x coordinate (In the frame of a series expansion and the use of the first element.), [14]. Therefore the approximative boundary condition on the basis of the contour function

$$F(x, r) = f(x) - \frac{l}{\mathcal{R}}r = 0\,, \tag{7.29}$$

with $\mathcal{R}$ the maximum thickness radius and l a reference length is formulated by

$$\frac{\partial \varphi}{\partial r} = u\,\frac{\mathcal{R}}{l}\,\frac{df(x)}{dx}\,, \tag{7.30}$$

As for the two dimensional plane flow we contemplate two different flows past a specific family of body contours and ask for the similarity rules. We list below the perturbation equation, the boundary condition and the approximate relation for the pressure coefficient for the first flow

$$(1 - M_{\infty 1}^2)\frac{\partial^2 \varphi_1}{\partial x_1^2} + \frac{\partial^2 \varphi_1}{\partial r_1^2} + \frac{1}{r_1}\frac{\partial \varphi_1}{\partial r_1} = 0\,, \tag{7.31}$$

$$\frac{\partial \varphi_1}{\partial r_1} = u_1\,\frac{\mathcal{R}_1}{l_1}\,\frac{df(x_1)}{dx_1}\,, \tag{7.32}$$

$$c_{p1} = -\frac{2}{u_1}\frac{\partial \varphi_1}{\partial x_1} - \frac{1}{u_1^2}\left(\frac{\partial \varphi_1}{\partial r_1}\right)^2\,, \tag{7.33}$$

and for the second flow

$$(1 - M_{\infty 2}^2)\frac{\partial^2 \varphi_2}{\partial x_2^2} + \frac{\partial^2 \varphi_2}{\partial r_2^2} + \frac{1}{r_2}\frac{\partial \varphi_2}{\partial r_2} = 0\,, \tag{7.34}$$

$$\frac{\partial \varphi_2}{\partial r_2} = u_2 \frac{\mathcal{R}_2}{l_2} \frac{df(x_2)}{dx_2} , \tag{7.35}$$

$$c_{p2} = -\frac{2}{u_2}\frac{\partial \varphi_2}{\partial x_2} - \frac{1}{u_2^2}\left(\frac{\partial \varphi_2}{\partial r_2}\right)^2 , \tag{7.36}$$

where the equation for the pressure coefficient differs from that of the two dimensional plane flow, Eq. (7.13), by an additional term.

Applying the transformation between x_1, r_1, φ_1 and x_2, r_2, φ_2

$$x_2 = x_1 , \qquad r_2 = a r_1 , \qquad \varphi_1(x_1, r_1) = C_r\, \varphi_2(x_2, r_2) , \tag{7.37}$$

one obtains with Eq. (7.31)

$$(1 - M_{\infty 1}^2)\frac{\partial^2 \varphi_2}{\partial x_2^2} + a^2 \frac{\partial^2 \varphi_2}{\partial r_2^2} + \frac{a^2}{r_1}\frac{\partial \varphi_2}{\partial r_2} = 0 . \tag{7.38}$$

This agrees with Eq. (7.34) only if

$$a^2 = \frac{1 - M_{\infty 1}^2}{1 - M_{\infty 2}^2} .$$

The boundary conditions Eqs. (7.32) and (7.35) leads to

$$\frac{d\, f(x_1)}{dx_1} = \frac{l_1}{\mathcal{R}_1 u_1}\frac{\partial \varphi_1}{\partial r_1} = \frac{l_1}{\mathcal{R}_1 u_1} a C_r \frac{\partial \varphi_2}{\partial r_2} = \frac{d\, f(x_2)}{dx_2} = \frac{l_2}{\mathcal{R}_2 u_2}\frac{\partial \varphi_2}{\partial r_2} , \tag{7.39}$$

and

$$C_r = \frac{1}{a}\frac{\mathcal{R}_1}{\mathcal{R}_2}\frac{l_2}{l_1}\frac{u_1}{u_2} . \tag{7.40}$$

From the pressure coefficients Eqs. (7.33) and (7.36) together with the relations for a and C_r one finds

$$c_{p1} = -\frac{2}{u_2}\frac{\partial \varphi_2}{\partial x_2}\left(\frac{1}{a}\frac{\mathcal{R}_1}{l_1}\frac{l_2}{\mathcal{R}_2}\right) - \frac{1}{u_2^2}\left(\frac{\partial \varphi_2}{\partial r_2}\right)^2\left(\frac{\mathcal{R}_1}{l_1}\frac{l_2}{\mathcal{R}_2}\right)^2 . \tag{7.41}$$

With the family of body contours, which satisfies the following thickness ratio[7]

[7]This ratio inserted in Eq. (7.40) leads to $C_r = \left(\frac{\mathcal{R}_1}{l_1} \Big/ \frac{\mathcal{R}_2}{l_2}\right)^2 \frac{u_1}{u_2}$.

$$\frac{\mathcal{R}_1}{l_1} \Big/ \frac{\mathcal{R}_2}{l_2} = \sqrt{\frac{1 - M_{\infty 2}^2}{1 - M_{\infty 1}^2}} ,$$

the pressure coefficient transforms[8] to

$$c_{p1} = \frac{1 - M_{\infty 2}^2}{1 - M_{\infty 1}^2} \left\{ -\frac{2}{u_2} \frac{\partial \varphi_2}{\partial x_2} - \frac{1}{u_2^2} \left(\frac{\partial \varphi_2}{\partial r_2} \right)^2 \right\} ,$$

$$c_{p1} = \frac{1 - M_{\infty 2}^2}{1 - M_{\infty 1}^2} c_{p2} , \tag{7.42}$$

which characterizes a streamline analogy and is similar to Eq. (7.23) (Göthert rule), for more details see also [7, 17].

7.5 Similarity Rules for Two and Three Dimensional Transonic Flows Past Slender Bodies (C)

The potential equation for small perturbations of the basic flow for transonic Mach numbers ($1 - \varepsilon \leqq M_\infty \leqq 1 + \varepsilon,\ \varepsilon = \mathcal{O}(0.1)$) is non-linear and reads

$$(1 - M_\infty^2)\, \varphi_{xx} + \varphi_{yy} + \varphi_{zz} = (\gamma + 1) M_\infty^2 \, \frac{1}{u_\infty} \, \varphi_x \, \varphi_{xx} . \tag{7.43}$$

This equation and the related similarity considerations were first derived and published by G. Guderley, Th. v. Kármán and K. Oswatitsch [8, 18, 19]. Details to the derivation of the above equation can also be found in [20]. The transonic approach to Mach number 1 can be performed from

- the subsonic side $M_\infty \lessapprox 1$. This means that for a freestream Mach number lower than 1 the flow around a given configuration, for example a thin wing, a thin profile or a blunt body like a sphere generates due to the flow expansion along the body contour supersonic pockets, which are usually terminated by a shock wave, Fig. 7.6. Further, if the freestream Mach number approaches unity, theoretically a perpenticular Mach wave is created at the upstream infinity, which travels quickly downstream for $M_\infty = 1 + \delta M (\delta M \ll 1)$ and gains more and more in strength composing a shock wave with discontinuous flow conditions across it.
- the supersonic side $M_\infty \gtrapprox 1$. In general a detached shock wave is generated, where behind it a subsonic flow regime exists. Due to the flow expansion around the body contour further downstream supersonic Mach numbers are again achieved. Thereby the subsonic and the supersonic regimes are separated by the sonic line.

[8] We mention that all these relations are valid also for supersonic Mach numbers, where instead of $1 - M_\infty^2$ the absolute value $|1 - M_\infty^2|$ has to be used.

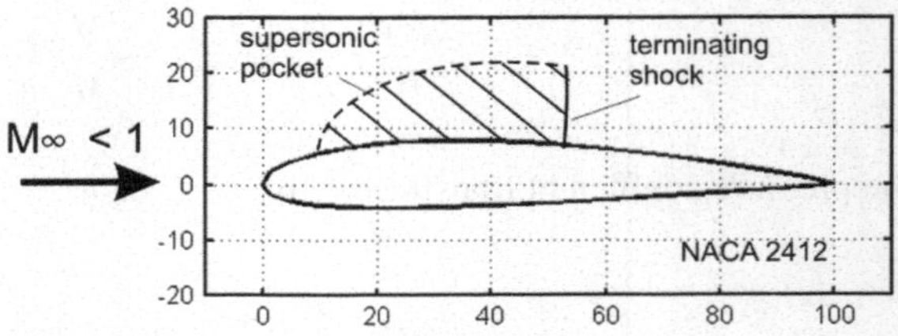

Fig. 7.6 Sketch of a subsonic/transonic flow situation past a NACA profile

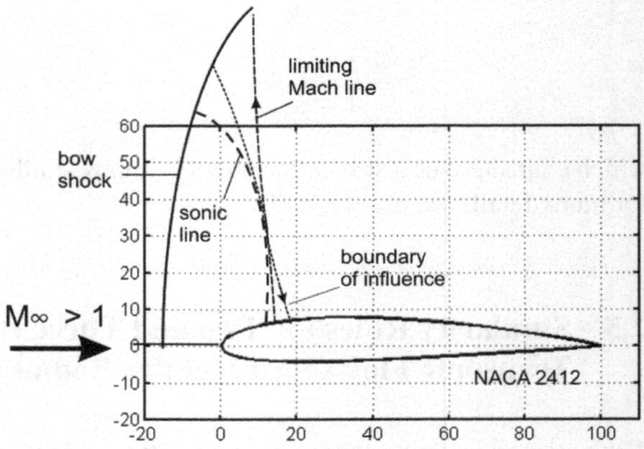

Fig. 7.7 Sketch of a supersonic/transonic flow situation past a NACA profile

In Fig. 7.7 the left running limiting Mach line is drawn (which attaches the sonic line) along this the body contour has its last opportunity to influence the subsonic region. On the other hand the right running Mach line coming from the bow shock delimitates the supersonic region which can be influenced by changes within the subsonic regime, Fig. 7.7.

The above described behavior of transonic/supersonic flows past blunt bodies (sphere or hemisphere-cylinder) can be observed and studied by the nearly exact solution of the Euler equations, given, e.g., in [21, 22]. The left part of Fig. 7.8 shows the numerical result of the flow past a hemisphere-cylinder for $M_\infty = 1.05$, with the contours of the bow shock, the sonic line and the left and right running Mach lines (characteristics, Mach waves).[9] Probably the only fully converged solution of the flow past a sphere with the very-close-to-sonic Mach number $M_\infty = 1.02$ is presented in the right part of Fig. 7.8. This figure shows the bow shock, the sonic line and the computational grid applied. It should be mentioned that for all the transonic supersonic flow computations a bow shock fitting technique was employed, which guarantees the best possible accuracy for the bow shock prediction.

Such solutions provide the scientist with all the fluiddynamic variables at every point of the flow field. In particular the stand-off distance of the bow shock and the pressure distribution along the centerline up to the stagnation point can be extracted from the solution, Fig. 7.9.

---

[9]Details regarding the solution and the comparison with other investigations can be found in [21].

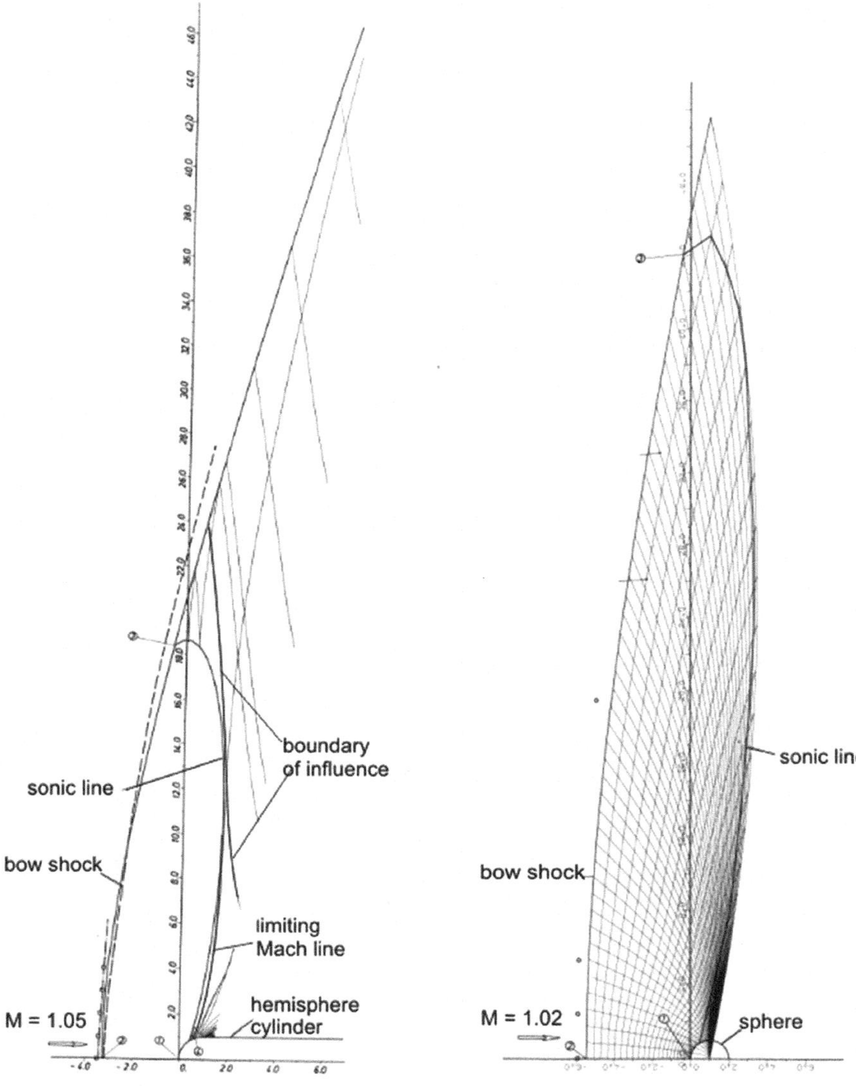

Fig. 7.8 High accuracy numerical solutions of the Euler equations for Mach numbers very close to unity. More details regarding the solutions can be found in [21, 22]

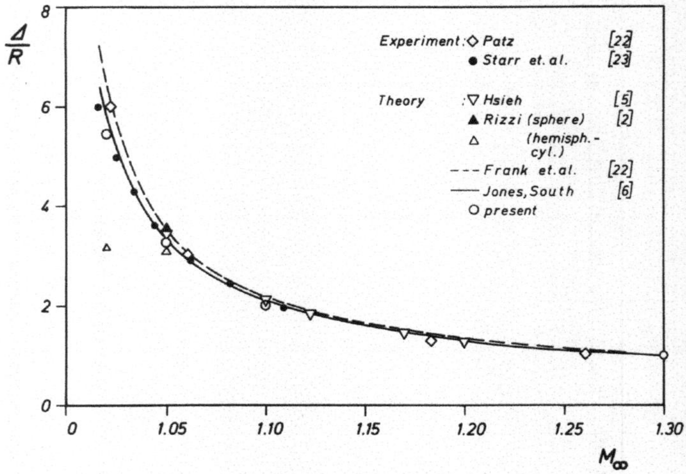

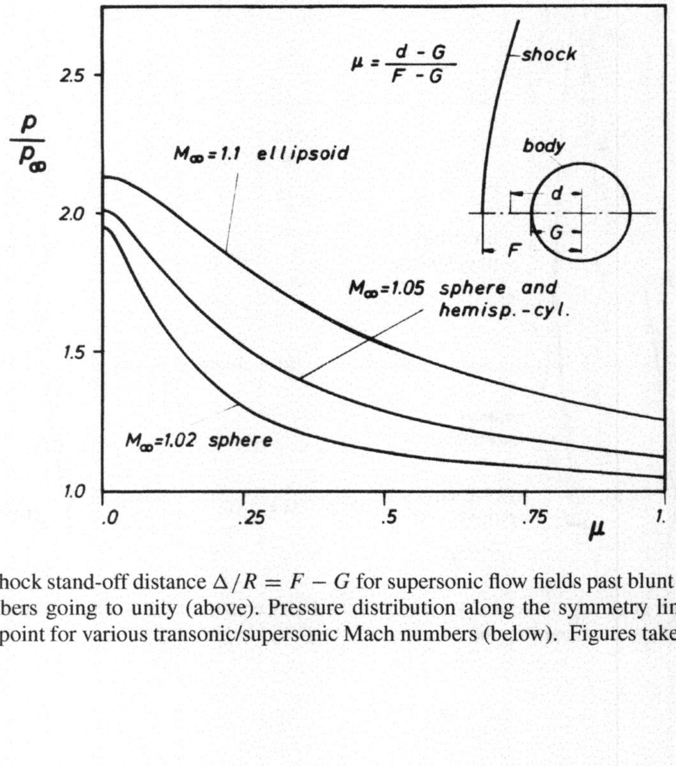

Fig. 7.9 Shock stand-off distance $\Delta/R = F - G$ for supersonic flow fields past blunt bodies with Mach numbers going to unity (above). Pressure distribution along the symmetry line up to the stagnation point for various transonic/supersonic Mach numbers (below). Figures taken from [21]

Applying the procedure of Sect. 7.3 to the transonic small perturbation potential equation (Eq. (7.43)) for thin bodies and considering the similarity transformations the non-linear term cares for an additional parameter, which has to be satisfied.

This quantity is called the transonic similarity parameter and reads

$$\chi = \frac{M_\infty^2 - 1}{\left(\frac{d}{l}(\gamma + 1)M_\infty^2 \frac{1}{u_\infty}\right)^{\frac{2}{3}}}, \tag{7.44}$$

which has the consequence that the Mach number cannot be changed independently of the thickness ratio d/l and that a change of the span width with the Mach number directly changes also the thickness ratio, which is in contrast to the results regarding the similarity rules of the linear small perturbation equation. For more information and details see [16].

In conclusion one can learn from the consideration of the similarity rules for flows past slender bodies at small angles of attack some qualitative behavior of the flow fields, but quantitatively, in particular today, the value of these considerations, for example for the work of design engineers is restricted.

7.6 Similarity Rules for Two and Three Dimensional Hypersonic Flows Using Linearized and Non-linearized Approximations (D)

During the time period from the 1920s to the 1960s the theory of hypersonic flows has been developed. The original equations, which were considered are the steady Euler equations, (Eq. (7.1)), describing inviscid rotational flow fields. This means that no viscous influence is acting and that the entropy is constant along streamlines (if they do not pass any shock wave), but not perpendicular to these. The physical content of the Euler equations for steady flow is also given by the following three equations: the gasdynamical basic equation, Crocco's law and the algebraic energy equation, which were used for the elaboration of the hypersonic theory, [23, 24]:

$$(u^2 - a^2)\frac{\partial u}{\partial x} + (v^2 - a^2)\frac{\partial v}{\partial y} + (w^2 - a^2)\frac{\partial w}{\partial z} +$$
$$vw\left(\frac{\partial v}{\partial z} + \frac{\partial w}{\partial y}\right) + uv\left(\frac{\partial u}{\partial y} + \frac{\partial v}{\partial x}\right) + uw\left(\frac{\partial u}{\partial z} + \frac{\partial w}{\partial x}\right), \tag{7.45}$$

$$rot\, \mathbf{v} \times \mathbf{v} = T\, grad\, s\,, \tag{7.46}$$

$$c_p T + \frac{|\mathbf{v}|^2}{2} = const\ . \tag{7.47}$$

Of course, the first condition for hypersonic flows is $M_\infty \gg 1$. Since the set of Eqs. (7.45)–(7.47) could not be solved analytically it was necessary to simplify these equations to find approximate solutions. This has been done in the same sense as described in Sects. 7.3 and 7.5. Considering flow fields where all perturbations of the flow velocity are small compared to the speed of sound, which means at least to examine slender bodies, the relation holds

$$\frac{v}{a} = \frac{|\mathbf{v}|}{a}\frac{v}{|\mathbf{v}|} = M \sin\vartheta \approx M\vartheta\ , \tag{7.48}$$

with v the component of the flow vector in y-direction, $\mathbf{v}$ the flow vector, a the speed of sound, M the local Mach number and ϑ the flow angle. Equation (7.48) exhibits that the slenderness of a configuration ($\vartheta \ll 1$) is not sufficient for the linearization of the governing equations as long as $v/a \ll 1$ is considered. Instead for very high Mach numbers the quantity $K = M\vartheta$ has to be looked at. K is called the Tsien parameter, [10]. The hypersonic flow theory deals with the outcome of the conditions

$$\text{(a)}\quad K = M_\infty\vartheta \ll 1 \qquad \text{and} \quad \text{(b)}\quad K = M_\infty\vartheta \gg 1$$

For the case $K \ll 1$ it can be deduced that for example the pressure coefficient[10] c_p has the same form as in the linearized theory, Eq. (7.13).

Remark Equation (7.13) reads

$$c_p = -\frac{2}{u}\frac{\partial\varphi}{\partial x}, \quad \frac{\partial\varphi}{\partial x} = u - u_\infty = \delta u$$

with u the velocity in x - direction. From the theory of linearized supersonic flow one finds with $|\mathbf{v}|^2 = u^2 + v^2 \quad \Rightarrow \quad |\mathbf{v}|\delta\mathbf{v} = u\delta u$ with $\delta v = 0$ and $a = u$ the speed of sound. Then the relation holds, [1, 13]

$$\frac{\delta|\mathbf{v}|}{|\mathbf{v}|} = -\frac{\vartheta}{\sqrt{M^2-1}}\ ,$$

and therefore in analogy to Eq. (7.13) we obtain

$$c_p = -\frac{2\vartheta}{\sqrt{M^2-1}}, \quad \vartheta << 1.$$

As the relation for the pressure coefficient has yielded the method to handle flows regarding $K \ll 1$, $\vartheta \ll 1$ is identical to those of the standard supersonic flows,

---

[10]This is also valid for the aerodynamic coefficients, like the lift C_L or the drag C_D.

Sect. 7.3. On the other hand, for $K \gg 1$ one speaks about the "genuine hypersonic flow", where no linearizations of the governing equations are feasible. This includes therefore also blunt body flows ($\vartheta \gg 1$).

We now consider a high Mach number ($M_\infty \gg 1$) flow around a thin flat plate at a small angle of attack $\alpha \ll 1$ (which corresponds to a small turning angle ϑ), Fig. 7.10. For the expansion at the upper side by using the Prandtl–Meyer relation, which is reduced by the above conditions, the pressure coefficient reads [12, 23]

$$\frac{c_p}{\alpha^2} = \frac{2}{\gamma (M_\infty \alpha)^2}\left(1 - \left(1 + \frac{\gamma - 1}{2} M_\infty \alpha\right)^{\frac{2\gamma}{\gamma-1}}\right) . \tag{7.49}$$

From the shock relations again for small angles of attack (originally for small turning angles ϑ and shock angles Θ) as well as high Mach numbers one finds for the pressure coefficient

$$\frac{c_p}{\alpha^2} = \frac{\gamma + 1}{2} + \sqrt{\left(\frac{\gamma + 1}{2}\right)^2 + \frac{4}{(M_\infty \alpha)^2}} . \tag{7.50}$$

The lift coefficient C_L of the inclined thin flat plate then[11] is

$$C_L = \alpha^2 \left[\frac{2}{\gamma (M_\infty \alpha)^2}\left(1 - \left(1 - \frac{\gamma - 1}{2} M_\infty \alpha\right)^{\frac{2\gamma}{\gamma-1}}\right) + \frac{\gamma + 1}{2} + \sqrt{\left(\frac{\gamma + 1}{2}\right)^2 + \frac{4}{(M_\infty \alpha)^2}}\right] . \tag{7.51}$$

The expansion part of Eq. (7.51) is only valid for $M_\infty \alpha \leqq 2/(\gamma - 1)$, because for $M_\infty \alpha = 2/(\gamma - 1)$ vacuum is reached at the upper side of the plate. We mention that for the inclined flat plate the drag coefficient is related to the lift coefficient by

$$\frac{C_L}{\alpha^2} = \frac{C_D}{\alpha^3} .$$

A consideration of the Eqs. (7.45)–(7.47) in the hypersonic limit $M_\infty \Rightarrow \infty$ was first performed by K. Oswatitsch [11] and has led to a system of equations, in which the freestream Mach number M_∞ does no longer explicitly appear (see for details [12, 17]). This means that the solution of the system is independent of the freestream Mach number, which is valid for all configurations including blunt bodies, [24]. Therefore a general property of hypersonic flow is given by the Mach number independence

[11] The negative sign in the term $\left(1 - \frac{\gamma-1}{2} M_\infty \alpha\right)$ is due to the counterclockwise definition of the angle of attack α.

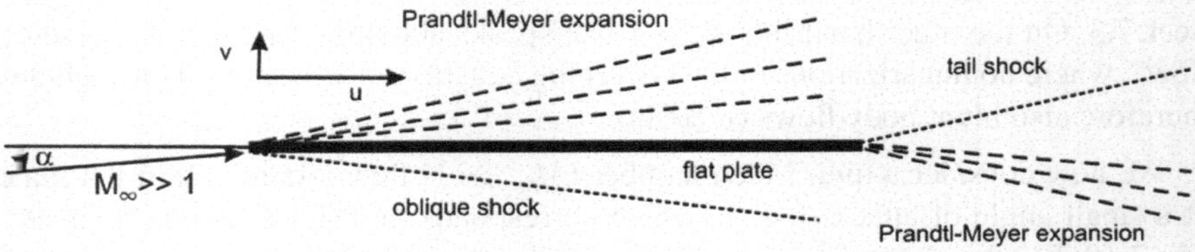

Fig. 7.10 Sketch of a supersonic-hypersonic flow past an inclined thin flat plate

principle. This encompasses the overall flow pattern with the bow shock, the sonic line, the Mach lines, the shock standoff distance and the streamlines and all the aerodynamic coefficients including the pressure one, but not the pressure and the entropy itself. For demonstration of this behavior Fig. 7.11 shows a series of flow fields past a blunt body (sphere-ogive) where the flow pattern obviously converges to a constant state already for a Mach number of $M_\infty \approx 4$.

When we restrict our investigations further to slender bodies with a thickness ratio $\tau \ll 1$—which is theoretically equivalent to a small turning angle ϑ of the flow or a small angle of attack α—the similarity rules for hypersonic flow around such configurations can be formulated as follows, [12]:

- the velocity component in x-direction transforms with $\dfrac{u}{u_\infty} - 1 \sim \tau^2$,
- the velocity component in y-direction transforms with $\dfrac{v}{u_\infty} \sim \tau$,
- the pressure coefficient transforms by $\dfrac{c_p}{\tau^2} = f(x, y/\tau, K, \gamma)$,

or one can state that two hypersonic flows past slender bodies of different slenderness τ_1 and τ_2, when K and γ are constant, are similar, if the relations

$$\begin{aligned}
\frac{1}{\tau^2}\left(\frac{u}{u_\infty} - 1\right)\bigg|_1 &= \frac{1}{\tau^2}\left(\frac{u}{u_\infty} - 1\right)\bigg|_2 ,\\
\frac{1}{\tau}\left(\frac{v}{u_\infty}\right)\bigg|_1 &= \frac{1}{\tau}\left(\frac{v}{u_\infty}\right)\bigg|_2 ,\\
\frac{y}{\tau}\,\Big|_1 &= \frac{y}{\tau}\,\Big|_2 ,\\
\frac{c_p}{\tau^2}\,\Big|_1 &= \frac{c_p}{\tau^2}\,\Big|_2
\end{aligned} \tag{7.52}$$

hold.

All the theoretical/analytical considerations regarding flow in the various Mach number regimes should have the goal to improve the understanding and the prediction of the aerodynamics around real configurations. In this regard we discuss below the supersonic and hypersonic behavior of several space vehicles, where their

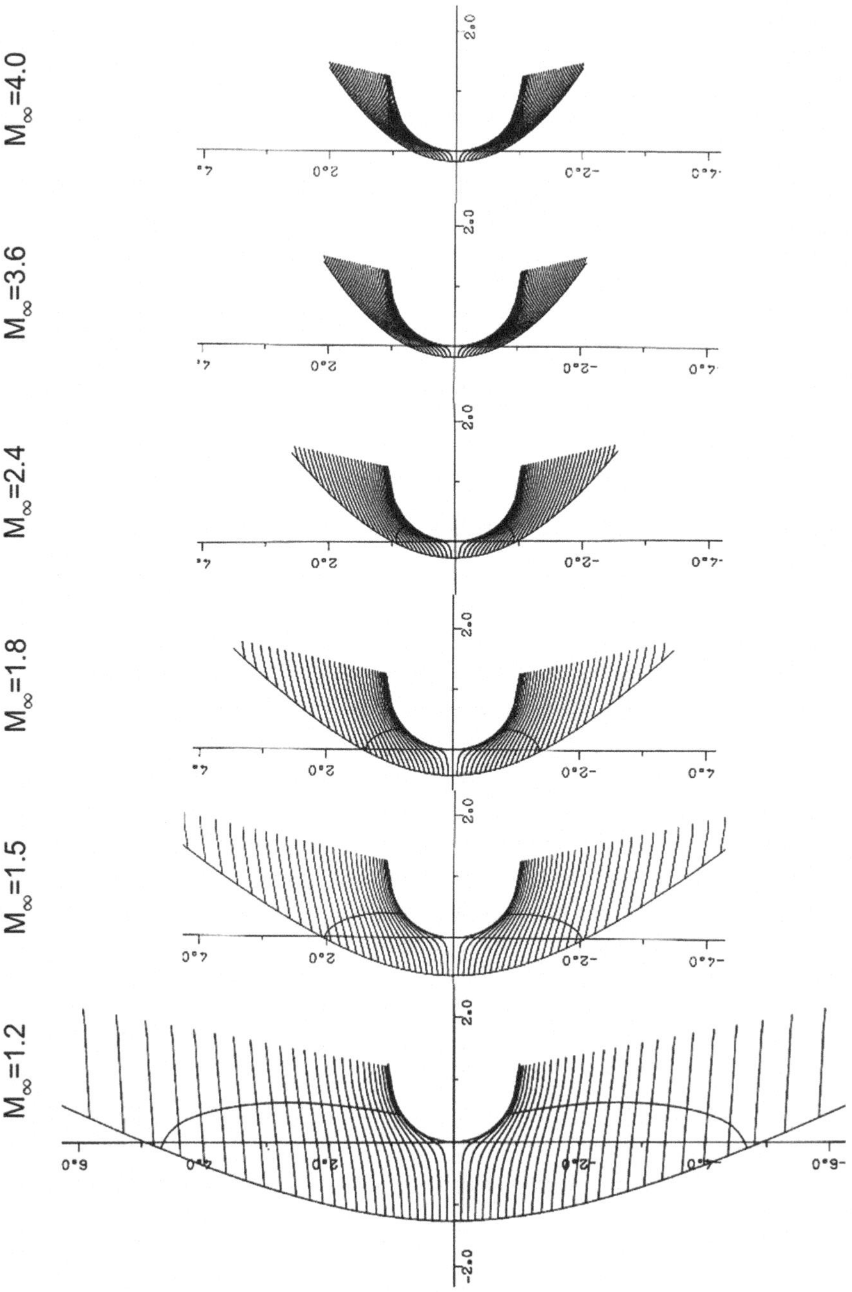

Fig. 7.11 Flow past a sphere-ogive configuration with increasing freestream Mach numbers showing the independence of flow characteristics (bow shock, sonic line, shock standoff distance, streamline pattern, etc.) with increasing Mach number ⇒ Oswatitsch's hypersonic independence principle, [11]. Flow fields computed with the method reported in [21, 22]

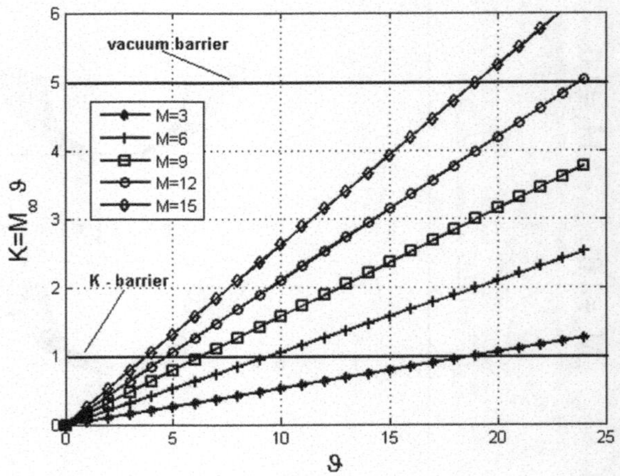

Fig. 7.12 Tsien parameter K as function of the turning angle ϑ. The K - barrier ($K = 1$) separates the linear from the non-linear hypersonic theory. For the inclined thin flat plate problem expansion is possible up to the vacuum—barrier ($M_\infty \alpha = 2/(\gamma - 1) = 5.$ for $\gamma = 1.4$), see Eq. (7.51)

aerodynamics were established by extended experimental, numerical and free flight investigations, [25].

7.7 The Aerodynamics of Real Space Vehicles in the Light of Supersonic and Hypersonic Approximate Theories

In Sect. 7.6 the Tsien parameter $K = M_\infty \vartheta$ was introduced. For $K < 1$ the linearized theory can be utilized, Sect. 7.3, whereas for $K > 1$ the "genuine hypersonic flow" is existing, which has to be described by non-linear relations. When we have a look at Fig. 7.12, the linearized theory can be applied up to $M_\infty \approx 3$ for angles of attack $\alpha \approx 20°$ ($\vartheta \Rightarrow \alpha$), whereas for $M_\infty \gtrapprox 12$ it holds only for small angles of attack ($\alpha \approx 4°$). It is interesting to find this behavior also in Fig. 7.13, where Ackeret's formula and Eq. (7.51) are plotted. For $M_\infty \approx 3$ both relations provide more or less identical results up to $\alpha \approx 15°$. For higher Mach numbers we observe for low α values linearity between both C_L curves, but this disappears for increasing α and the non-linear character of Eq. (7.51) dominates.

7.7.1 Similarities of Aerodynamic Data at Hypersonic Mach Numbers

The monograph, [25], presents the aerodynamic data (as far as they are available) of most of the space vehicles, which were considered in the past in development programs, feasibility studies and real projects. We investigate the lift and drag coef-

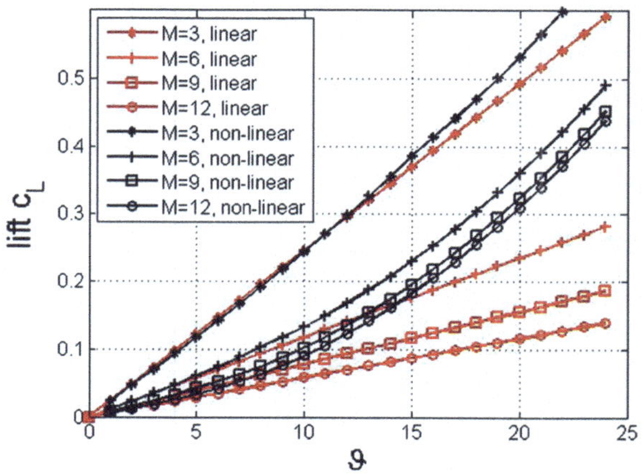

Fig. 7.13 Lift coefficient of the inclined thin flat plate as function of turning angle ϑ ($\Rightarrow$ angle of attack α). The linear values are due to Ackeret's formula $4\vartheta/\sqrt{M_\infty^2 - 1}$, whereas the non-linear values are based on Eq. (7.51)

ficients for supersonic and hypersonic Mach numbers (which we extract from [25]) for the winged re-entry vehicles (WRV) Shuttle Orbiter and Hermes[12] as well as the lifting re-entry vehicles (LRV) X-33 and X-38 as well as the Sänger hypersonic two-stage space transportation system with respect to the supersonic/hypersonic theory discussed in the sections above, [26, 27]. The first four planes are pressure dominated re-entry vehicles, whereas the lower stage of the Sänger configuration was a hypersonic cruise vehicle, where viscous effects play a major role, [28].

The Lift Coefficient

The lift coefficient C_L versus the angle of attack in the range $-10° \leqq \alpha \leqq 50°$ and for the two Mach numbers[13] $M_\infty = 8$ and 10 is plotted for four re-entry vehicles in Fig. 7.14. A first inspection of the curves exhibits that Oswatitsch's Mach number independence principle, Sect. 7.6, for hypersonic flows is well confirmed. A second observation reveals that all the curves have a common characteristic:

They start ($0 \lessapprox \alpha \lessapprox 30°$) with a weak quadratic non-linearity, experience then around $\alpha \approx 30°$ an inflection point[14] and merge into a sinusoidal behavior ($30° \lessapprox \alpha \lessapprox 50°$).

All aerodynamic coefficients are individually made non-dimensional by a reference area S_{ref}. Typically S_{ref} should reflect a characteristic area. For a re-entry vehicle this could be approximately a projection of the configuration into the $x - y$ plane. Unfortunately this is not always the case. For example, the reference area of the Shuttle Orbiter ($S_{ref} = 249.9$ m^2) is much smaller than the projected area, which is

[12] Later we extent the work to the Phoenix and X-34 configurations.

[13] For the X-38 configuration no data for $M_\infty = 8$ are available, instead the values of $M_\infty = 5.47$ are used, which obviously meet also the Mach number independence principle.

[14] The inflection point is necessary because for $\alpha \Rightarrow 90°$ the normal force turns over into the drag force and the axial force into the lift force which tends to zero.

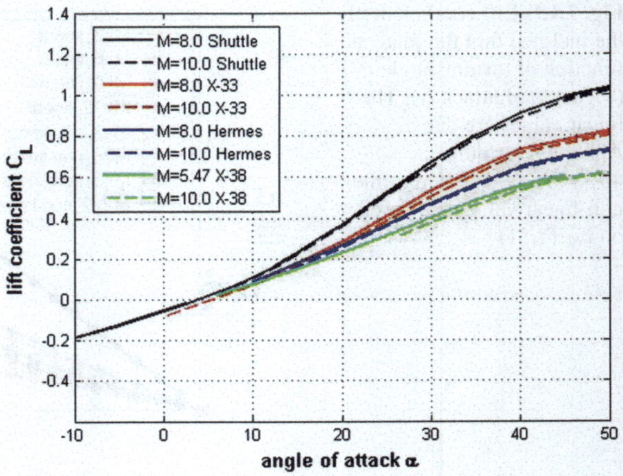

Fig. 7.14 Lift coefficients C_L for hypersonic Mach numbers as function of the angle of attack α for the WRV and LRV space vehicles. Data taken from the aerodynamic data bases given in [25]. The reference areas S_{ref} applied are individually defined in the frame of the specific space vehicle projects

approximately $S_{proj} \approx 350\ \text{m}^2$. For the Hermes vehicle we notice an inverse assignment, $S_{ref} = 84.7\ \text{m}^2$ to $S_{proj} \approx 80.7\ \text{m}^2$. To make the data comparable we therefore introduce reference area quantities S^*_{ref} which homogenize the non-dimensionality of the aerodynamic data. The S^*_{ref} values are determined by the following procedure:

The lift coefficient is defined by $C_L = \dfrac{L}{\rho u_\infty^2 S_{ref}}$. As an example the C_L values in the hypersonic flight regime of the Shuttle Orbiter and Hermes are considered. The idea then is to assume that $C_L(\alpha)|_{Shuttle} = a \cdot C_L(\alpha)|_{Hermes}$ applies for all angles of attack α. This leads to $\left.\dfrac{L}{S_{ref}}\right|_{Shuttle} = \left.\dfrac{L}{S_{ref}/a}\right|_{Hermes}$ and $S^*_{ref}|_{Hermes} = S_{ref}/a|_{Hermes}$ with a being the scaling factor.

The such determined S^*_{ref} values are listed in Table 7.1. When these quantities are applied to the C_L values of Fig. 7.14 the surprising result is received, that all the curves more or less coincide entirely, Fig. 7.15. This is all the more astonishing because the configurations of the re-entry vehicles have substantially different geometries, Fig. 7.16. The two-stage hypersonic space transportation system Sänger[15] is presented in Fig. 7.17. It is well known that in hypersonic flight already at moderate angles of attack α the upper surface of a vehicle does not play a role for the aerodynamic forces, because this area is in a near vacuum state (hypersonic shadow). Since the lower sides of the vehicles examined here are only moderately cambered, it make sense to compare the C_L values with the results of Eq. (7.51) for the inclined flat plate. In Fig. 7.18 these data are recorded. Of course the graph of Eq. (7.51) offers a nearly quadratic non-linearity in α, whereas the re-entry vehicle data exhibit a weaker non-linearity behavior. But nevertheless the hypersonic behavior of the vehicles are,

[15]Only the aerodynamics of the lower stage is considered.

Table 7.1 List of original reference areas S_{ref} and scaled reference areas S^*_{ref}

Vehicle	S_{ref} [m^2]	S^*_{ref} [m^2]	Scaling factor a
Shuttle	249.9	249.9	1
Hermes	84.7	60.2	1.408
X-33	149.4	117.8	1.268
X-34	33.21	35.33	0.9399
X-38	21.6	13.2	1.646
Phoenix	–	–	1.459
Sänger	1658	1131.7	1.465

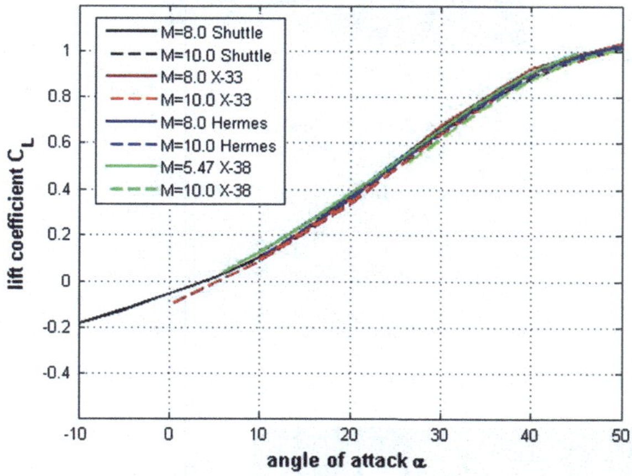

Fig. 7.15 Lift coefficients C_L at hypersonic Mach numbers as function of the angle of attack α for the WRV and LRV space vehicles. Data taken from the aerodynamic data bases given in [25]. In contrast to the diagram Fig. 7.14 the reference areas S^*_{ref} are used

for the angle of attack regime[16] $0° \lessapprox \alpha \lessapprox 30°$, not far from that of the inclined flat plate.

The Drag Coefficient

We turn now our interest to the drag coefficient C_D. Figure 7.19 exhibits the original data with S_{ref} of [25]. Again Oswatitsch's Mach number independence principle is impressively satisfied. Scaling the drag data with the reference areas S^*_{ref} leads to the diagram Fig. 7.20. We recognize that in this case two groups of curves coincide nearly completely, namely these of the WRV group Shuttle Orbiter and Hermes as well as these of the LRV group X-33 and X-38.

For the hypersonic lower stage cruise vehicle Sänger, Fig. 7.17, the aerodynamic data are available up to Mach numbers $M_\infty = 7$, [25]. It is known that the Mach number independence principle is approached for blunt bodies, like the re-entry vehicles, already at $M_\infty \gtrapprox 4$, whereas for slender configurations with a pointed

[16]Beyond $\alpha \approx 30°$ the graph of Eq. (7.51) follows further its quadratic non-linearity and has no inflection point as the vehicle data have.

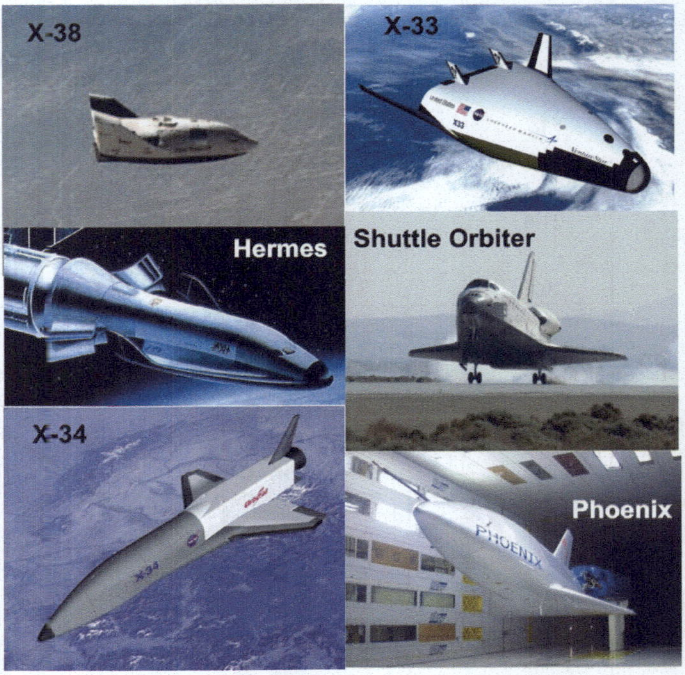

Fig. 7.16 Space vehicles: LRV: X-38, X-33; WRV: Hermes, Shuttle Orbiter, X-34, Phoenix

Fig. 7.17 The two-stage hypersonic space transportation system Sänger

nose, like the hypersonic cruise vehicle Sänger, this occurs in principle at higher Mach numbers. Nevertheless the lift values for $M_\infty = 7$ are also plotted together with the graphs of Fig. 7.15, which exhibit the same trend like the re-entry vehicles have, namely the weakly non-linear behavior, Fig. 7.21 (left). In contrast to that the drag values, plotted in Fig. 7.21 (right), show no conformity with the re-entry vehicle data.

The Pitching Moment Coefficient

In order to complete the longitudinal aerodynamic data base the pitching moment coefficient C_m as function of the Mach number and the angle of attack has to be considered. The definition of the position of the moment reference point by the respective designers of the various space vehicles is arbitrary to a certain extent.

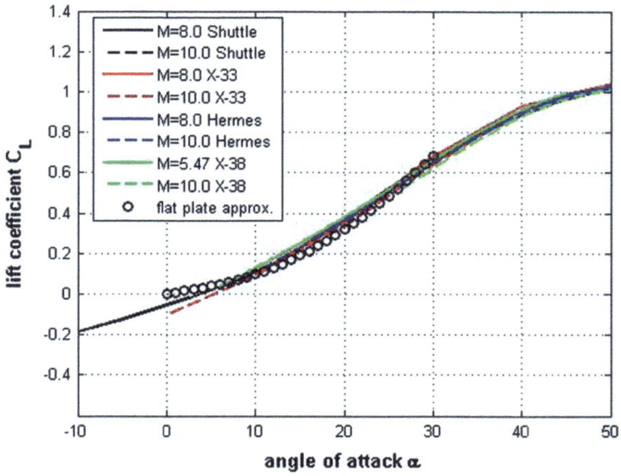

Fig. 7.18 C_L graphs of Fig. 7.15 and in addition the C_L curve for the inclined flat plate, Eq. (7.51)

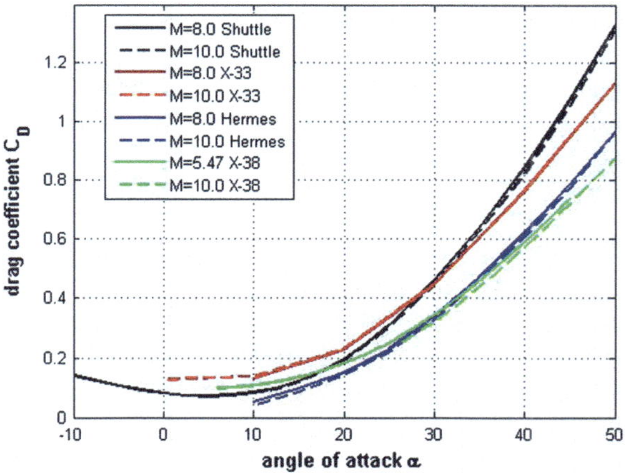

Fig. 7.19 Drag coefficients C_D for hypersonic Mach numbers as function of the angle of attack α for the WRV and LRV space vehicles. Data taken from the aerodynamic data bases given in [25]. The reference areas S_{ref} applied are individually defined in the frame of the specific space vehicle projects

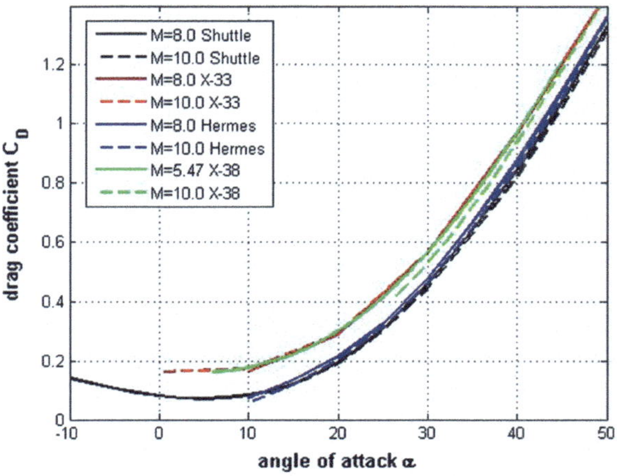

Fig. 7.20 Drag coefficients C_D for hypersonic Mach numbers as function of the angle of attack α for the WRV and LRV space vehicles. Data taken from the aerodynamic data bases given in [25]. In contrast to the diagram Fig. 7.19 the reference areas are homogenized with S^*_{ref}

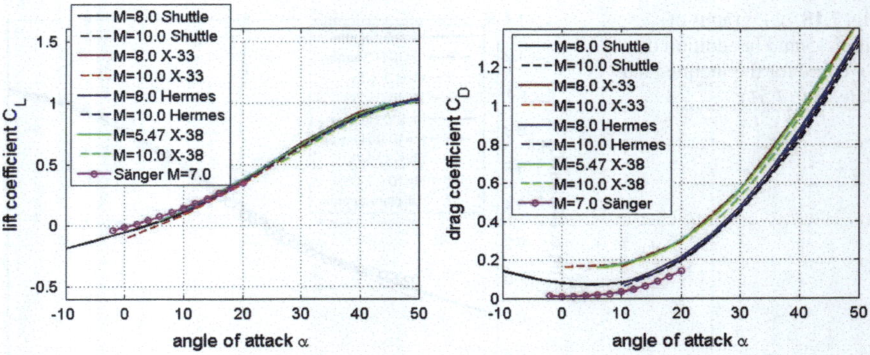

Fig. 7.21 Sänger data with $M_\infty = 7$ are included in diagram Fig. 7.15 for the C_L values, left figure, and in diagram Fig. 7.20 for the C_D values, right figure

Therefore the comparison of the pitching moment coefficients related to the defined reference points makes no sense. A better situation exists, if the moment reference points are transformed to a common geometrical point, for example the point in the nose with $x^*, z^* = 0$, see Fig. 7.22. The equation for the transformation then is given by

$$C_m(x^*, z^*) = C_m(x_{ref}, z_{ref}) + C_N(x^* - x_{ref}) - C_A(z^* - z_{ref}) , \tag{7.53}$$

with C_N being the normal force coefficient and C_A the axial force coefficient. Typically for flight vehicles, which are in a certain sense slender, like civil and combat aircraft or winged and lifting re-entry vehicles as well as hypersonic spacecraft the influence of the axial force $(C_A(z^* - z_{ref}))$ on the pitching moment is rather low due to the small shift of the z-coordinate and the low value of C_A compared to C_N. Therefore this term is omitted in the following considerations. For capsules and all sorts of cones, the situation is different which is described in detail in Chap. 7 of [29].

Since for the evaluation of Eq. (7.53) we need the normal force coefficient C_N, and in the general case also the axial force coefficient C_A, we provide these quantities[17] from the lift and the drag coefficients by

$$\begin{aligned} C_A &= C_D \cos\alpha - C_L \sin\alpha , \\ C_N &= C_D \sin\alpha + C_L \cos\alpha . \end{aligned} \tag{7.54}$$

For the hypersonic Mach number $M = 10$ Fig. 7.23 (left) exhibits the normal force with S_{ref} for the WRV space vehicles Shuttle Orbiter, Hermes and in addition Phoenix as well as the LRV space vehicles X-38 and X-33. Figure 7.23 (right) shows the graphs

[17]For flights with yaw angle $\beta = 0°$.

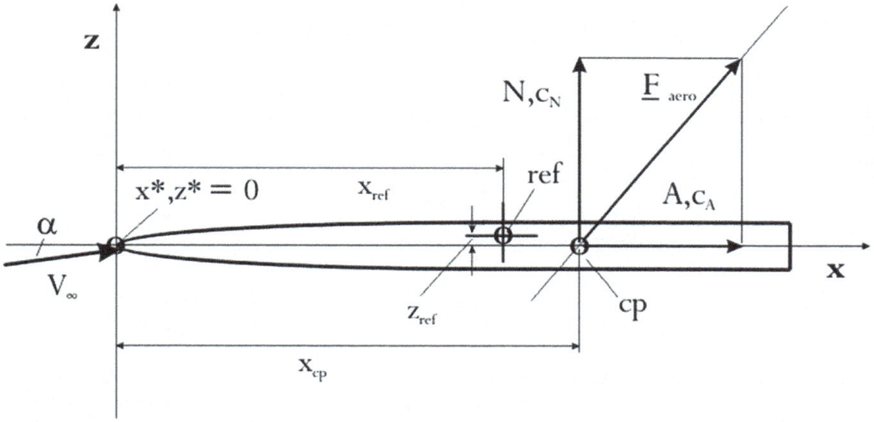

Fig. 7.22 Sketch of reference position x_{ref}, z_{ref} and transformed position x^*, z^* for the pitching moment evaluation. cp: center of pressure, ref: reference point

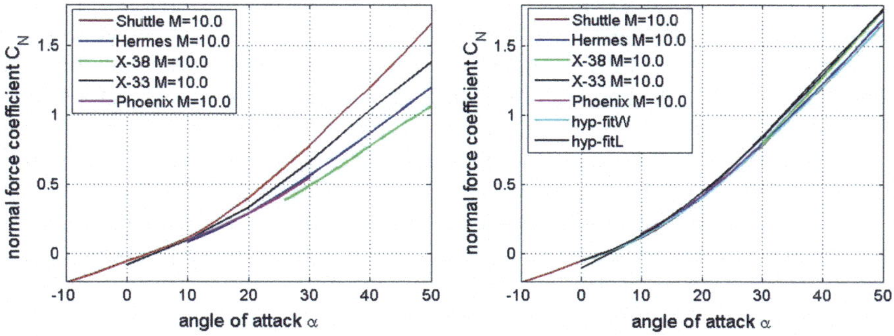

Fig. 7.23 Normal force coefficients C_N for $M = 10$ as function of angle of attack α for the WRV and LRV space vehicles. The graphs in the left figure are obtained with S_{ref} and the ones in the right figure with S^*_{ref}. The right figure contains also the normal force coefficients calculated with the semi-empirical formula Eq. (7.55)

scaled with S^*_{ref}. The figure contains also the results of the semi-empirical formula Eq. (7.55). The normal force coefficients behave very similar to the drag coefficients by forming also the WRV and LRV groups.

The axial force coefficients C_A are presented in Fig. 7.24. The left figure shows the graphs with S_{ref}. In the right figure the data plotted are scaled with S^*_{ref}, where also the curves evaluated by the semi-empirical formula Eq. (7.55) are included. Also in this case a grouping of the WRV as well as LRV space vehicles can be observed. There obviously exists the tendency to oscillate depending on the angle of attack, which is true also for the results of the semi-empirical formula.

An inspection of the data of the pitching moment, given in [25], shows that a certain ambiguity consists regarding the origin of the coordinate systems which are applied

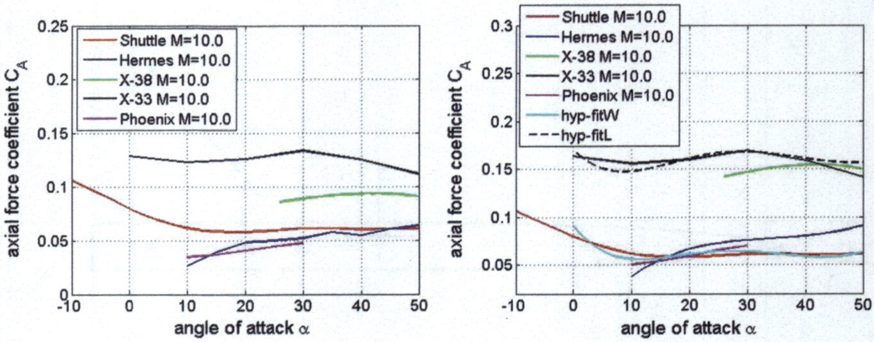

Fig. 7.24 Axial force coefficients for $M = 10$ as function of angle of attack α for the WRV and LRV space vehicles. The graphs in the left figure are obtained with S_{ref} and the ones in the right figure with S^*_{ref}. The right figure contains also the evaluations of the semi-empirical formula Eq. (7.55)

for the definition of the geometrical configuration and in particular the moment reference point. In other words it is often doubtful, if the given x_{ref} values have their origin in the nose of the configuration. Nevertheless we compare the pitching moment quantities of the above defined space vehicles when they are transformed to $x^* \Rightarrow 0$ by Eq. (7.53).

The x-component of the moment reference point x_{ref} of the space vehicles considered are listed in Table 7.2. When the pitching moment functions of the space vehicles are compared regarding the prescribed x_{ref} values, no commonality can be observed as is already argued above. Figure 7.25 (left) shows the graphs with respect to S_{ref} and Fig. 7.25 (right) the ones with respect to S^*_{ref}.

Transforming the pitching moment coefficients by Eq. (7.53) to $x^* = 0$ the picture changes due to the fact that the result of the transformation is normal-force dominated and exhibits therefore also a clear grouping, see Fig. 7.26. The graphs in the left figure are scaled with S_{ref} and the ones in the right figure with S^*_{ref}. From the experience gained by the consideration of the drag and the normal force behavior it is expected that the pitching moment coefficients scaled by S^*_{ref} show also the WRV and LRV grouping. But this is not the case.[18] The Hermes pitching moment coefficient, for example, departs somewhat from that of the Shuttle Orbiter.

As a consequence of the above argumentation, which encompasses the uncertainty from which points regarding the coordinate systems of the various space vehicles the x_{ref} values are measured, the transformation of the pitching moment to a nose position is changed from $x_{ref} = 0.60$ to $x'_{ref} = 0.66$ for the Hermes configuration and by the same reason from $x_{ref} = 0.66$ to $x'_{ref} = 0.58$ for the X-33 vehicle. The result is given in Fig. 7.27 and exhibits a more reasonable agreement between the data of the various vehicles.

---

[18]Obviously the Shuttle Orbiter, Phoenix and X-33 group together as well as Hermes and X-38, which is in contrast to the experience regarding the drag and normal forces.

Table 7.2 Coordinates x_{ref} of the moment reference points for the space vehicles considered

Space vehicle	Pitching moment reference point
Shuttle	$0.65 \cdot L_{ref}$
Hermes	$0.60 \cdot L_{ref}$
Phoenix	$0.68 \cdot L_{ref}$
X-38	$0.57 \cdot L_{ref}$
X-33	$0.66 \cdot L_{ref}$
X-34	$0.65 \cdot L_{ref}$

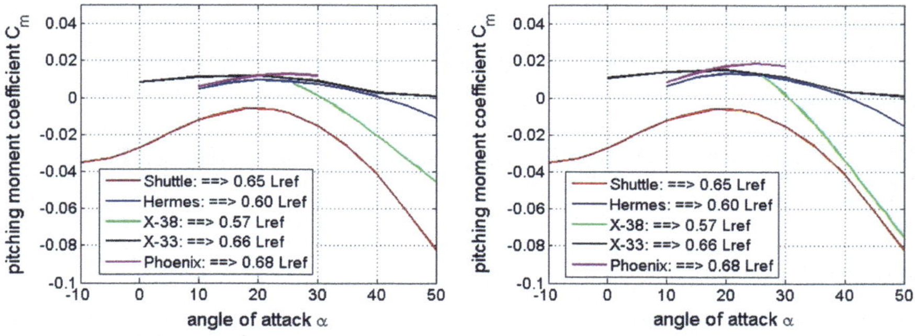

Fig. 7.25 Pitching moment coefficients C_m evaluated at the moment reference points x_{ref} for $M = 10$ for the WRV and LRV space vehicles. The graphs in the left figure are obtained with S_{ref} and the ones in the right figure with S^*_{ref}

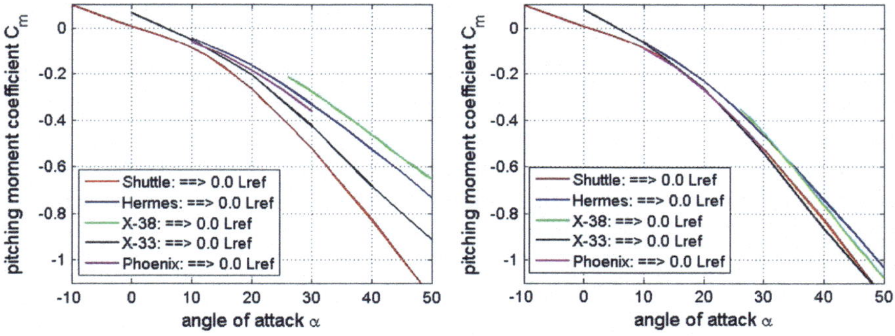

Fig. 7.26 Pitching moment coefficients C_m evaluated at the positions $x^* = 0$, see Fig. 7.22, for $M = 10$ for the WRV and LRV space vehicles. The graphs in the left figure are obtained with S_{ref} and the ones in the right figure with S^*_{ref}

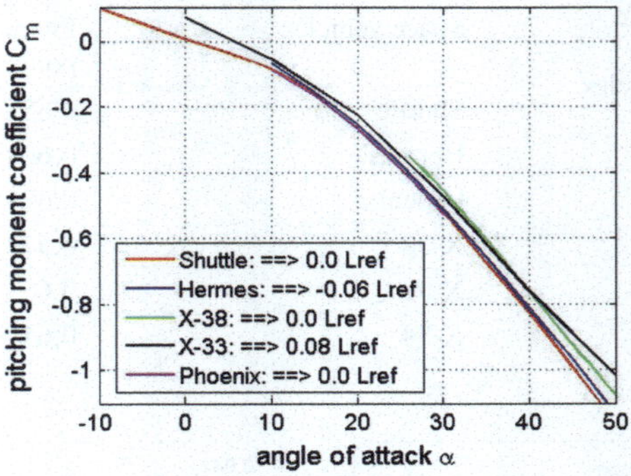

Fig. 7.27 Pitching moment coefficients C_m with adjusted moment reference points for Hermes ($x_{ref} = 0.60 \Rightarrow x'_{ref} = 0.66$) and X-33 ($x_{ref} = 0.66 \Rightarrow x'_{ref} = 0.58$)

7.7.2 Similarities of Aerodynamic Data for Supersonic Mach Numbers

The Lift Coefficient

For the supersonic regime the aerodynamic data of the three re-entry vehicles Shuttle Orbiter, Hermes and X-38[19] and the hypersonic lower stage cruise vehicle Sänger are evaluated for the three Mach numbers $M_\infty = 2, 3, 4$. The Mach number $M_\infty = 2$ plot for the lift coefficient C_L reveals that Hermes, X-38 and Sänger behaves nearly linear, whereas the Shuttle Orbiter show a weak non-linearity for $\alpha > 10°$, Fig. 7.28 (upper left). Scaling the data by the reference areas S^*_{ref}, as was done for the hypersonic cases, groups the graphs together exhibiting a more or less equal slope, Fig. 7.28 (upper right). The result of Eq. (7.51) for the inclined flat plate flow is also plotted. This curve shows a weak non-linearity in the sense that beyond $\alpha \gtrsim 15°$ it does not follow the space vehicle data.

For the Mach numbers[20] $M_\infty = 3$ and 4 a growing non-linearity can be observed with the surprising situation that the Shuttle Orbiter data fit nearly perfect with the results of Eq. (7.51), Fig. 7.28 (middle left and lower left). Scaling also these curves with the reference areas S^*_{ref} the C_L graphs exhibit a good coincidence, Fig. 7.28 (middle right and lower right), which has nearly the quality found in the hypersonic Mach number regime, Figs. 7.14 and 7.15. In particular it must be emphasized that the characteristics of the lift coefficient C_L of the re-entry vehicles as well as the hypersonic cruise vehicle are very similar.

The Drag Coefficient

The drag coefficient C_D for the three supersonic Mach numbers $M_\infty = 2, 3, 4$ and the space vehicles Shuttle Orbiter, Hermes, X-38 and Sänger is illustrated in the three left diagrams of Fig. 7.29. Scaling the data with S^*_{ref}, Table 7.1, the Shuttle Orbiter and

[19]Unfortunately supersonic data for the X-33 vehicle are not available.

[20]For X-38 no data are available for $M_\infty = 3$.

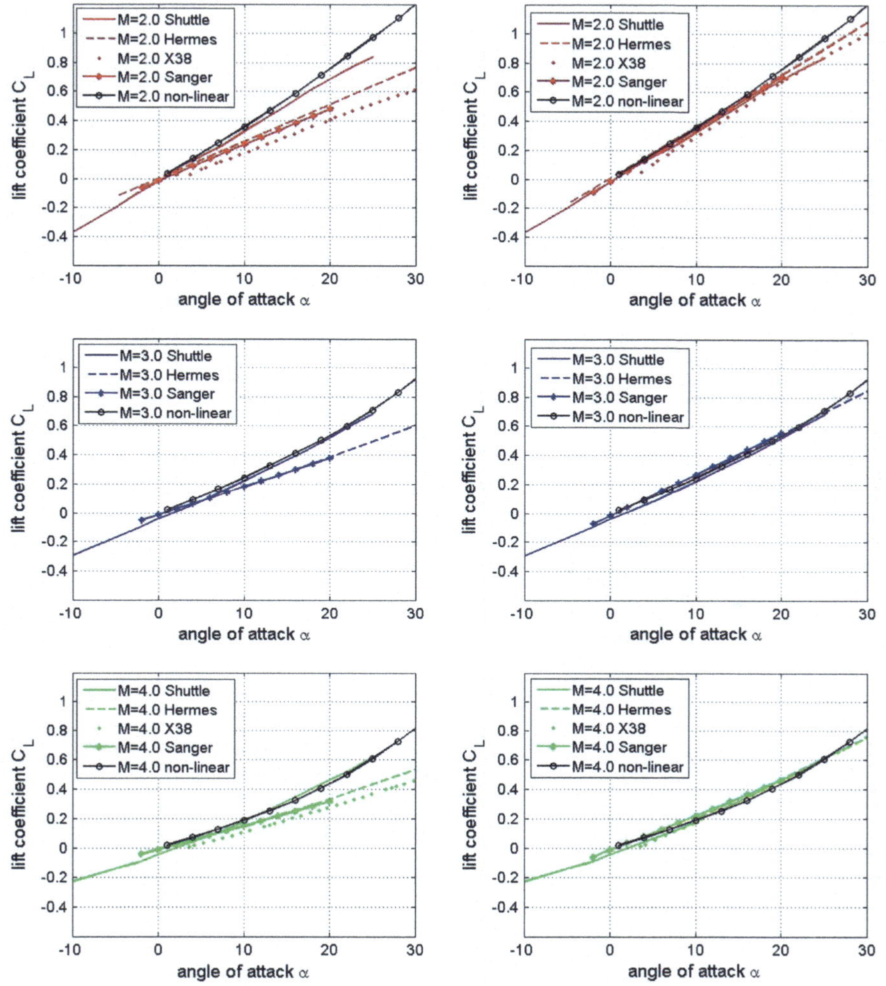

Fig. 7.28 Lift coefficients C_L for the space vehicles: Shuttle Orbiter, Hermes, X-38 and Sänger for $M_\infty = 2, 3, 4$. Included are also the results of Eq. (7.51) for the inclined flat plate

Hermes (WRV) graphs indicate again a certain conformity, but the X-38 and Sänger data reveal no commonality with the other graphs (see the three right diagrams of Fig. 7.29). The Sänger configuration is slender with a more or less pointed nose which is in contrast to the re-entry vehicles, which are substantially blunt. Therefore the disagreement of the Sänger data of the drag coefficient C_D compared to the other ones seems to be plausible, in particular because the Sänger flow field is more friction dominated than the flow fields of the re-entry vehicles. The evaluation of Eq. (7.51) for the inclined flat plate regarding the drag coefficient is also plotted in the right diagrams, Fig. 7.29, and indicates a fairly good agreement with the Sänger data.

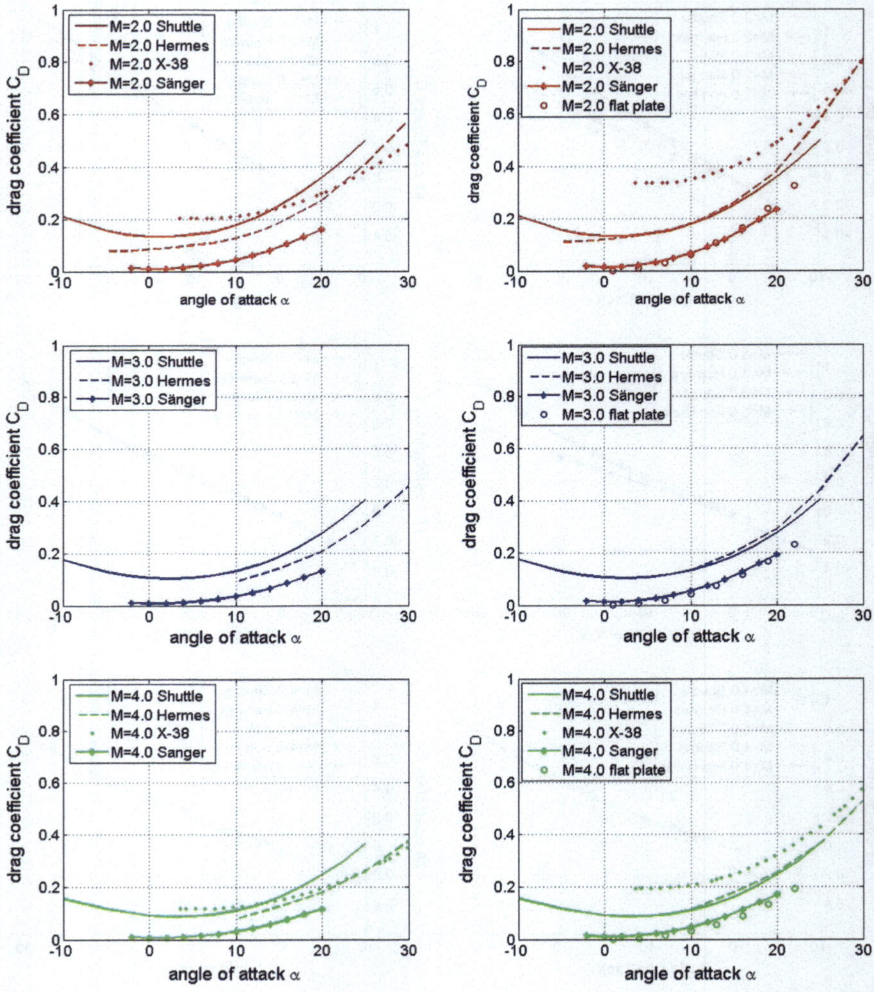

Fig. 7.29 Drag coefficients C_D for the space vehicles: Shuttle Orbiter, Hermes, X-38 and Sänger for $M_\infty = 2, 3, 4$. The left figures show the original data taken from [25]. In the right figures the data scaled by S^*_{ref} are plotted and additionally the drag data for the inclined flat plate evaluated by Eq. (7.51) are included

The Pitching Moment Coefficient

As an example for the supersonic Mach number regime the pitching moment coefficients for $M = 2$ are evaluated and compared. Since there are no data available for the X-33 vehicle for this Mach number, this vehicle is replaced by the X-34 space plane, [25]. As mentioned earlier the X-33 vehicle belongs to the LRV group, whereas the X-34 vehicle obviously is a member of the WRV group. Firstly, as was done for the hypersonic case, the normal and axial force coefficients are determined, which are

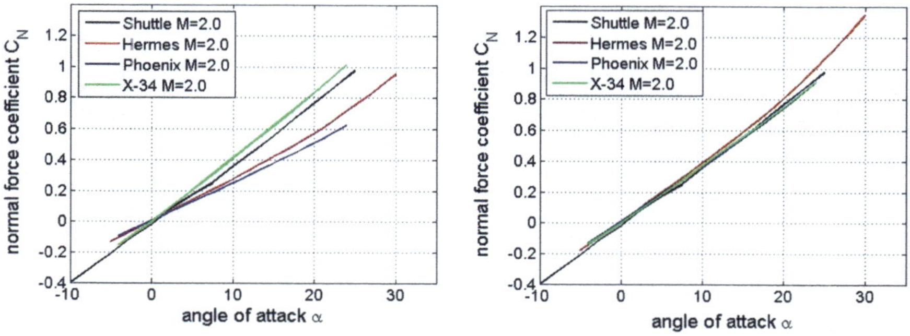

Fig. 7.30 Normal force coefficients C_N for $M = 2$ for the WRV space vehicles. The graphs in the left figure are obtained by S_{ref} and the ones in the right figure by S^*_{ref}

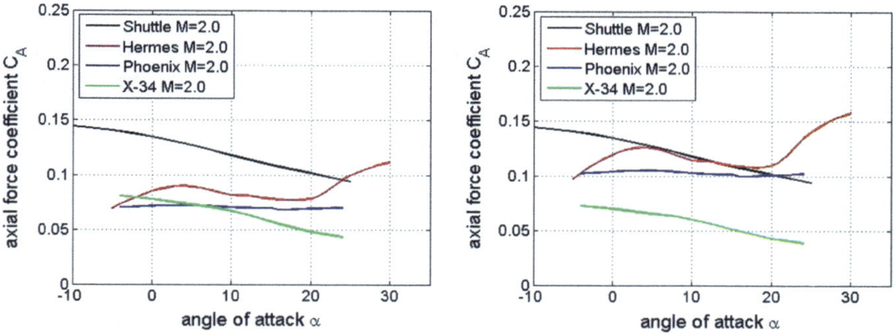

Fig. 7.31 Axial force coefficients C_A for $M = 2$ for the WRV space vehicles. The graphs in the left figure are obtained by S_{ref} and the ones in the right figure by S^*_{ref}

necessary for the transformation by Eq. (7.53). In Fig. 7.30 the normal force coefficients are plotted. The left figure shows the graphs scaled by S_{ref}. The slope and the position of the graphs of the Shuttle Orbiter, Hermes, X-34 and Phoenix vehicles, when scaled by S^*_{ref}, look very similar (right figure).

Figure 7.31 is devoted to the axial force coefficients. As it was already observed in the hypersonic flow regime, Fig. 7.24, the axial force coefficients show not so a clear grouping as the other aerodynamic coefficients have. Again in the left figure the scaling was done by S_{ref} and in the right figure by S^*_{ref}.

In the supersonic Mach number $M_\infty = 2$ case the pitching moments evaluated at the various x_{ref} positions have obviously no commonality, Fig. 7.32. The evaluation of the pitching moment coefficients at the $x^* = 0$ position can be found in Fig. 7.33. The graphs in the left figure are scaled by S_{ref} and in the right figure by S^*_{ref}. Generally the slope of all the curves in the right figure are nearly the same (within a certain

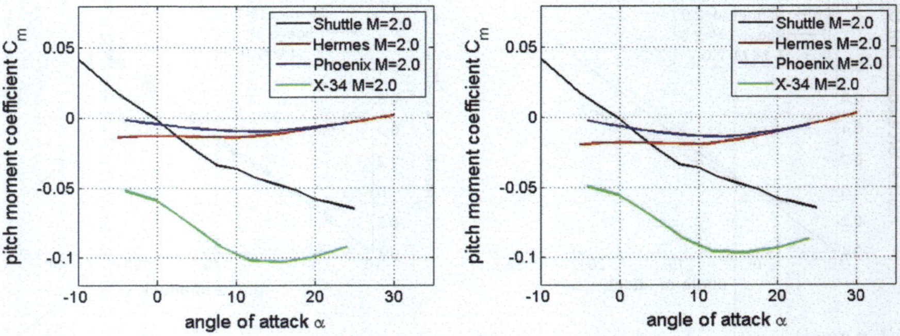

Fig. 7.32 Pitching moment coefficients C_m evaluated at the moment reference points x_{ref} for $M = 2$. The graphs in the left figure are obtained by S_{ref} and the ones in the right figure by S^*_{ref}

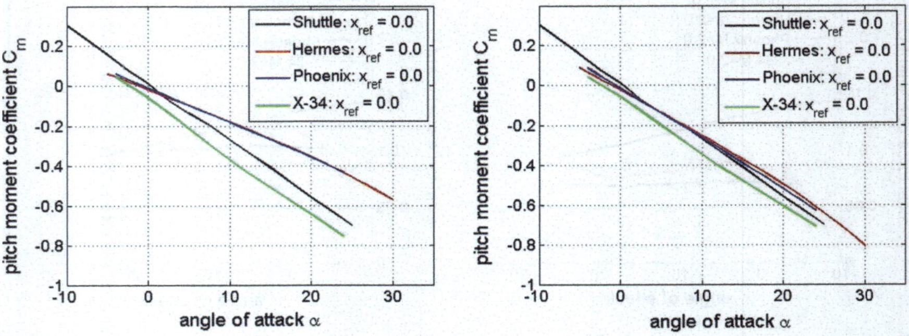

Fig. 7.33 Pitching moment coefficients C_m evaluated at the positions $x^* = 0.$, see Fig. 7.22, for $M = 2$. The graphs in the left figure are obtained by S_{ref} and the ones in the right figure by S^*_{ref}

bandwidth), but there exists a moderate shift of the X-34 data away from the position of the graphs of the Shuttle Orbiter, Hermes and Phoenix.[21]

7.7.3 Construction of a Fit Equation for the Aerodynamic Data at Hypersonic and Supersonic Mach Numbers

The following exponential function is able to approximate all the graphs with respect to lift, drag and pitching moment for hypersonic as well as supersonic Mach numbers as is discussed below:

[21]The shift can not be explained by the uncertainties about the origin of the coordinate systems and the corresponding x_{ref} values.

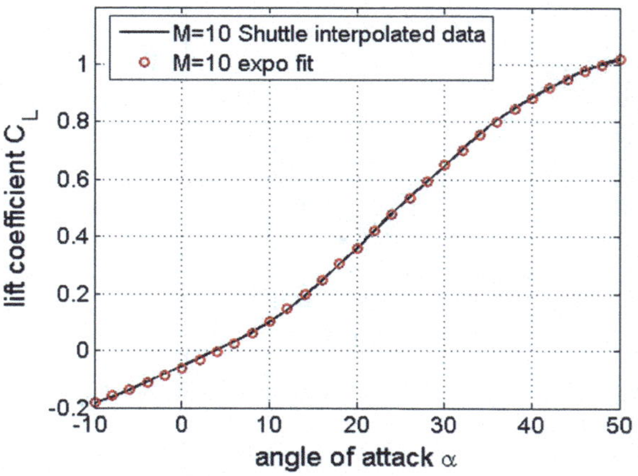

Fig. 7.34 Lift coefficient C_L for $M_\infty = 10$ for the Shuttle Orbiter. Comparison with the fit Eq. (7.55) using the coefficients $CLhyp$

$$C_i = a_0 + a_1 \alpha e^{-d_3 \alpha} + a_2 e^{-d_3 |d_2 \alpha|^{d_4}}, \qquad C_i \div C_N, C_A, C_L, C_D, C_m. \tag{7.55}$$

Fit Equation of Lift at Hypersonic Mach Numbers

There exists obviously in the hypersonic regime for all Mach numbers, due to Oswatitsch's Mach number independence principle, and for all re-entry configurations considered here for the lift coefficient C_L just one common function $C_L(\alpha)$, if the homogenized reference areas S^*_{ref} are used, see Table 7.1 and Fig. 7.15. This function is approximated by the semi-empirical formula Eq. (7.55) by using the coefficients $CLhyp$:

$$\begin{aligned} &d_2 = 1.15, \quad d_3 = 2.30, \quad d_4 = 2.15 \\ &a_0 = 1.0417, \quad a_1 = 0.7546, \quad a_2 = -1.1033, \qquad CLhyp. \end{aligned}$$

In Fig. 7.34 the Shuttle Orbiter data[22] for $M_\infty = 10$ are compared with the data of Eq. (7.55) using the coefficients $CLhyp$ and show a nearly perfect agreement. This means that the equation is also an excellent approximation for all the graphs of Fig. 7.15.

Fit Equation of Drag at Hypersonic Mach Numbers

The drag coefficients C_D for the WRV and LRV space vehicles, when scaled with the homogenized reference areas S^*_{ref}, coincide relatively good with each other, see

---

[22] Since the data originally are given only every 5° an interpolation has been performed in order to have the data available every 2°.

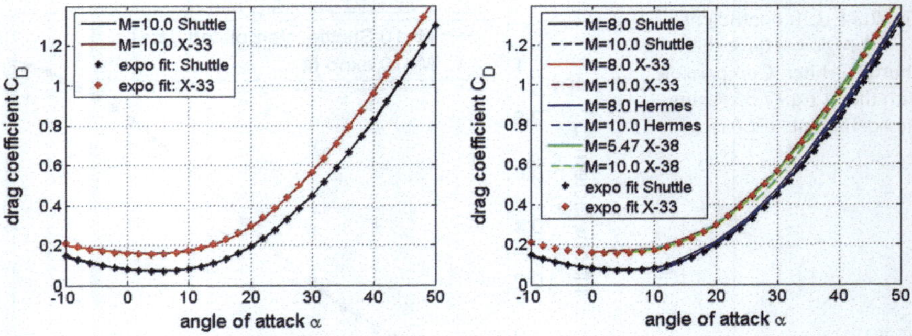

Fig. 7.35 Drag coefficients C_D for space vehicles at hypersonic flight. In the left figure the values with $M_\infty = 10$ for the Shuttle Orbiter and the X-33 vehicle are plotted which are compared with the fit equation, Eq. (7.55) using the coefficients $CDhypW$ and $CDhypL$. The right figure corresponds to Fig. 7.20 in which the results of Eq. (7.55) are included

Fig. 7.20. As for the lift coefficients above the fit Eq. (7.55) with the corresponding sets of coefficients for the WRV and LRV groups are applied.

The set of coefficients $CDhypW$ for the WRV group reads

$$\begin{aligned} &d_2 = 1.00, \quad d_3 = 0.72, \quad d_4 = 2.50 \\ &a_0 = 3.3460, \quad a_1 = -0.1922, \quad a_2 = -3.2662, \qquad CDhypW, \end{aligned}$$

and for the LRV group $CDhypL$

$$\begin{aligned} &d_2 = 1.00, \quad d_3 = 0.72, \quad d_4 = 2.50 \\ &a_0 = 2.7574, \quad a_1 = -0.0870, \quad a_2 = -2.6311, \qquad CDhypL. \end{aligned}$$

Figure 7.35 (left) shows for $M_\infty = 10$ the comparison between the fit Eq. (7.55) with either the coefficients $CDhypW$ or $CDhypL$ and the Shuttle Orbiter as well as the X-33 data. All the data of Fig. 7.20 supplemented by Eq. (7.55) can be found in Fig. 7.35 (right).

Fit Equation of Pitching Moment at Hypersonic Mach Numbers

It was shown in Fig. 7.26 that the pitching momentv coefficients, evaluated at $x_{ref} = x^* = 0$, group only to some extent together, but not in the sense of the WRV and LRV groups regarding the drag and normal forces. Nevertheless with the set of coefficients $Cmhyp$ given by

$$\begin{aligned} &d_2 = 0.78, \quad d_3 = 1.0, \quad d_4 = 2.16, \\ &a_0 = -2.8074, \quad a_1 = -0.5107, \quad a_2 = 2.8264, \qquad Cmhyp, \end{aligned}$$

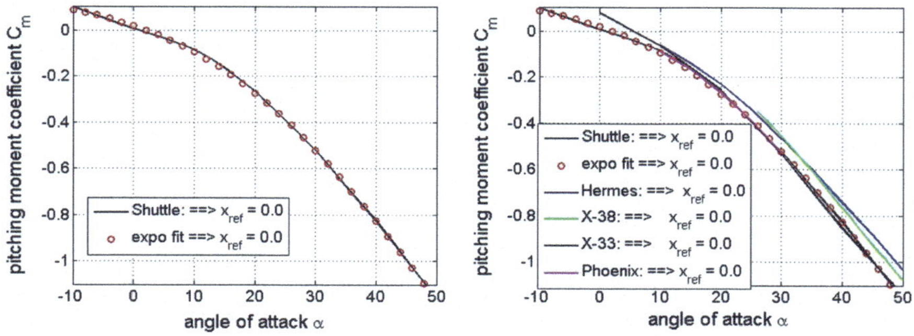

Fig. 7.36 Pitching moment coefficient C_m at $x_{ref} = x^* = 0$ for $M_\infty = 10$, see Fig. 7.26. In the left figure the Shuttle graph is compared with the result of Eq. 7.55 using $Cmhyp$. The right figure contains additionally the data of Hermes, X-38, X-33 and Phoenix

the fit Eq. (7.55) is applied to approximate the Shuttle Orbiter data. In Fig. 7.36 the graph of this equation is plotted together with the Shuttle data (left figure) and with all the data of Fig. 7.26, (right figure). This fit equation provides the engineer at least with a first estimate of the pitching moment coefficient of any new winged or lifting space vehicle.

An Instructive Example

For the Phoenix configuration the lift, drag and pitching moment coefficients at $M_\infty = 10$ are given in the angle of attack range $10° \leq \alpha \leq 30°$, see [25]. The angle of attack range should be extended to $0° \leq \alpha \leq 50°$. The extension is carried out by using the general fit equation (Eq. (7.55)) with $CLhyp$ for the lift, with $CDhypW$ for the drag and with $Cmhyp$ for the pitching moment (for $x^* = 0$), as well as the scaling factor $a = 1.459$ (or S^*_{ref}) for the Phoenix vehicle. The normal force coefficient is calculated by the lift and drag coefficients. The result is very convincing, see the diagrams in Fig. 7.37, where the approximated values are compared with the available Phoenix data, [25]. Lift, drag, normal force and pitching moment ($x^* = 0$) agree perfectly. The pitching moment approximation for $x_{ref} = 0.68\ L_{ref}$ (lower right figure) has an accuracy comparably to that in Fig. 7.25.

Fit Equation of Lift at Supersonic Mach Numbers

In Fig. 7.28 the lift coefficients regarding the Mach numbers $M_\infty = 2, 3, 4$ for the space vehicles: Shuttle Orbiter, Hermes, X-38 and Sänger are presented. A good agreement of the real project data, when scaled by S^*_{ref}, can be stated. For increasing Mach numbers the project data have developed a weak non-linearity. The data are compared with the non-linear formula Eq. (7.51) for a hypersonic ($M_\infty \gg 1$) inclined flat plate flow, which reveals a slightly stronger non-linearity than the project data.

Using also in this case the fit Eq. (7.55) with the coefficients

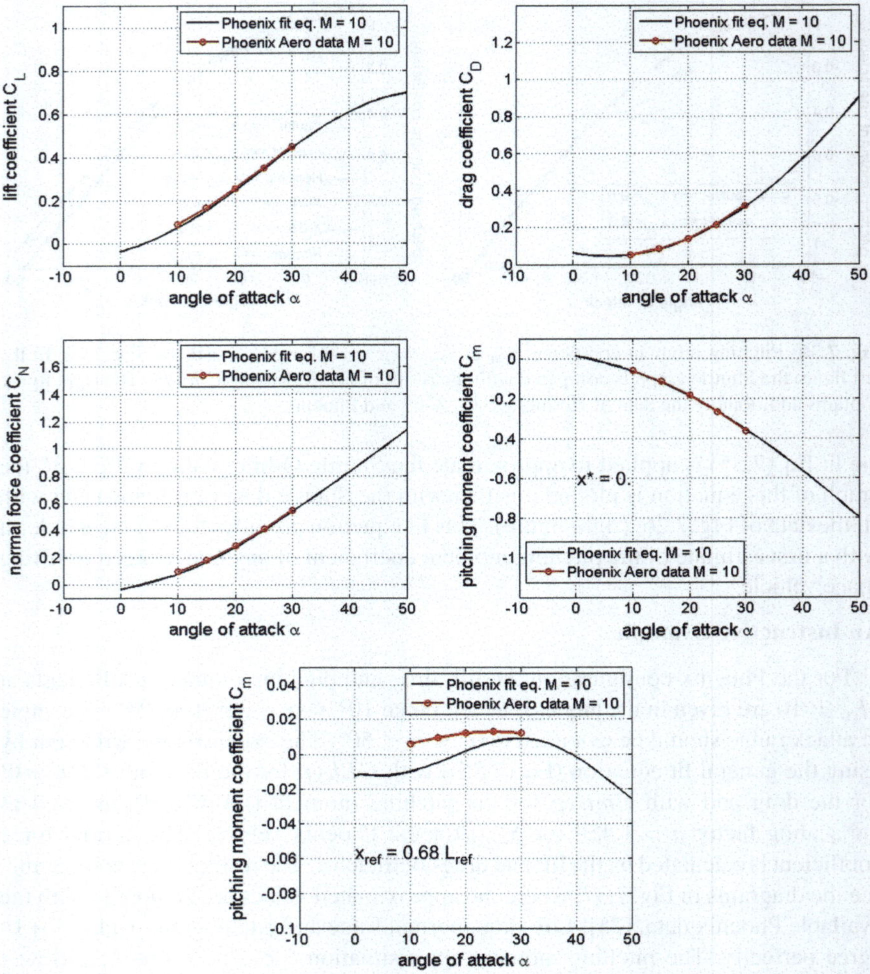

Fig. 7.37 Extension of the Phoenix longitudinal aerodynamics from the angle of attack range $10° \leq \alpha \leq 30°$ to $0° \leq \alpha \leq 50°$. Pitching moment coefficient ($x_{ref} = x^* = 0$) approximated by Eq. (7.55) with $Cmhyp$ (middle right). Pitching moment coefficient for $x_{ref} = 0.68L_{ref}$ computed by an inverse version of Eq. (7.53) (lower figure)

$$d_2 = 0.5, \;\; d_3 = 0.72, \;\; d_4 = 1.75\,,$$
$$a_0 = 4.7704, \;\; a_1 = 1.9992, \;\; a_2 = -4,7902, \;\; \text{for} \;\; M_\infty = 2, \qquad CLsup1\,,$$
$$d_2 = 0.5, \;\; d_3 = 0.72, \;\; d_4 = 2.10\,,$$
$$a_0 = 8.9337, \;\; a_1 = 1.4698, \;\; a_2 = -8.9747, \;\; \text{for} \;\; M_\infty = 3, \qquad CLsup2\,,$$
$$d_2 = 0.5, \;\; d_3 = 0.72, \;\; d_4 = 1.90\,,$$
$$a_0 = 7.4193, \;\; a_1 = 1.1703, \;\; a_2 = -7.4669, \;\; \text{for} \;\; M_\infty = 4, \qquad CLsup3\,.$$

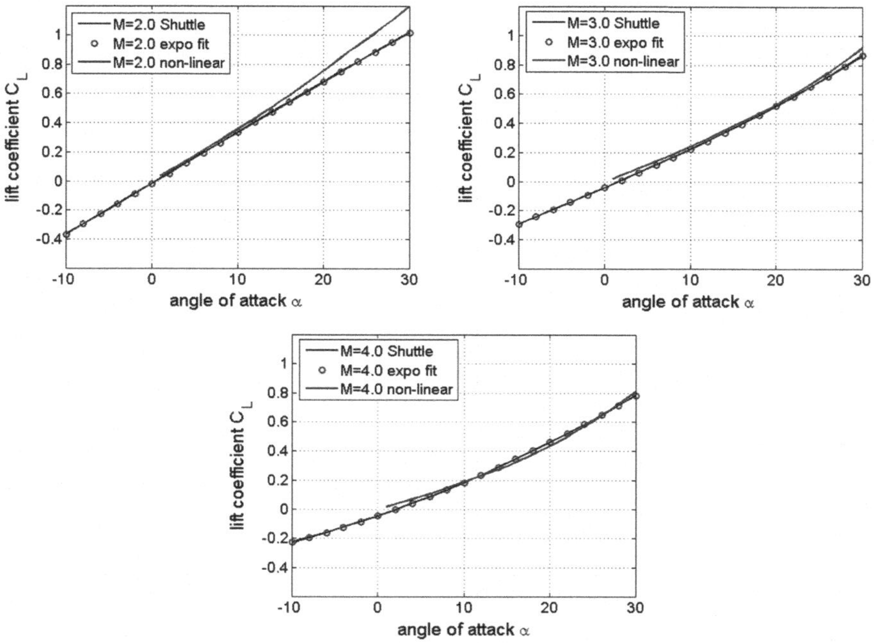

Fig. 7.38 Lift coefficients C_L in supersonic flight regime ($M_\infty = 2, 3, 4$). The above figures correspond to Fig. 7.28. Comparison of Shuttle Orbiter data with results of the non-linear Eq. (7.51) and the fit Eq. (7.55)

reflects a nearly perfect agreement, Fig. 7.38.

Fit Equation of Drag at Supersonic Mach Numbers

It make sense to have a closer look at the graphs of Fig. 7.29, where for the Mach numbers $M_\infty = 2, 3, 4$ and the space vehicles Shuttle Orbiter, Hermes, X-38 and Sänger the drag coefficients C_D are plotted. As was already noted, the data of the Shuttle Orbiter and Hermes form the WRV group. For $M_\infty = 2$, Fig. 7.39[23] (upper left), the agreement of both curves is satisfactory in the regime $0° \lesssim \alpha \lesssim 17°$, and is becoming better and better with increasing Mach number and is nearly perfect for $M_\infty = 4.$, $0° \lesssim \alpha \lesssim 25°$, Fig. 7.39 (upper right and lower middle). Totally different from these curves are the C_D data of the Sänger configuration, but they coincide surprisingly good with the evaluation of Eq. (7.51).

The data of the fit Eq. (7.55) with the coefficients

[23]The X-38 data are omitted in Fig. 7.39.

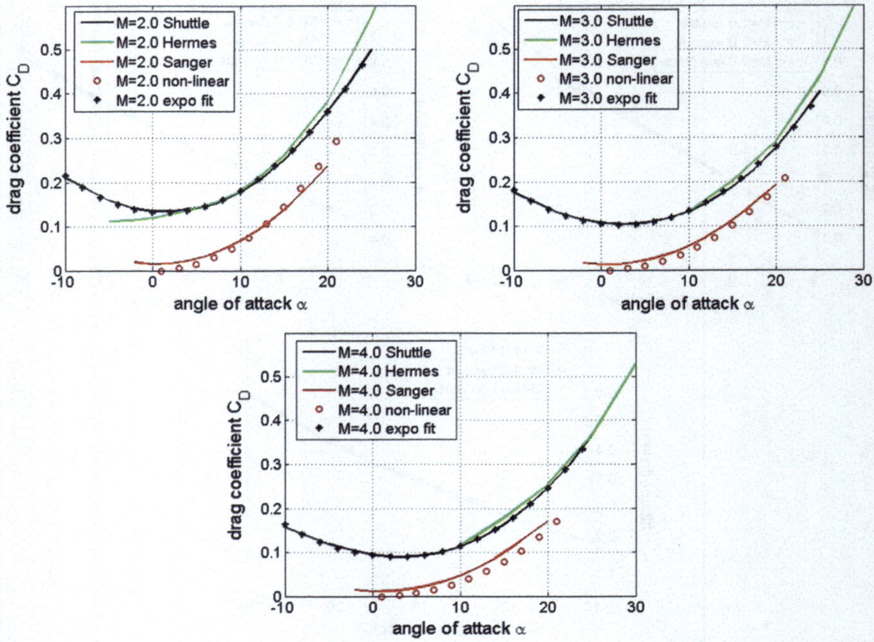

Fig. 7.39 Drag coefficients C_D in supersonic flight regime ($M_\infty = 2, 3, 4$) for the Shuttle Orbiter, Hermes (WRV group) and the Sänger configuration. The above figures correspond to Fig. 7.29. Included are also the results of Eqs. (7.51) and (7.55)

$$
\begin{aligned}
&d_2 = 0.5, \quad d_3 = 0.72, \quad d_4 = 2.00\,,\\
&a_0 = 11.8871, \quad a_1 = -0.1037, \quad a_2 = -11.7552, \quad \text{for } M_\infty = 2, \qquad CDsup1\,,\\
&d_2 = 0.5, \quad d_3 = 0.72, \quad d_4 = 2.10\,,\\
&a_0 = 11.7030, \quad a_1 = -0.1395, \quad a_2 = -11.5980, \quad \text{for } M_\infty = 3, \qquad CDsup2\,,\\
&d_2 = 0.5, \quad d_3 = 0.72, \quad d_4 = 2.20\,,\\
&a_0 = 12.6368, \quad a_1 = -0.1453, \quad a_2 = -12.5434, \quad \text{for } M_\infty = 4, \qquad CDsup3\,.
\end{aligned}
$$

complete the graphs in Fig. 7.39.

Fit Equation of Pitching Moment at Supersonic Mach Numbers

The pitching moment graph for the Shuttle with $M_\infty = 2$ is given in Fig. 7.33. The fit equation (7.55) with the coefficients $Cmsup$

$$
\begin{aligned}
&d_2 = 0.78, \quad d_3 = 1.0, \quad d_4 = 2.16,\\
&a_0 = -2.6689, \quad a_1 = -1.6375, \quad a_2 = 2.6745, \qquad Cmsup,
\end{aligned}
$$

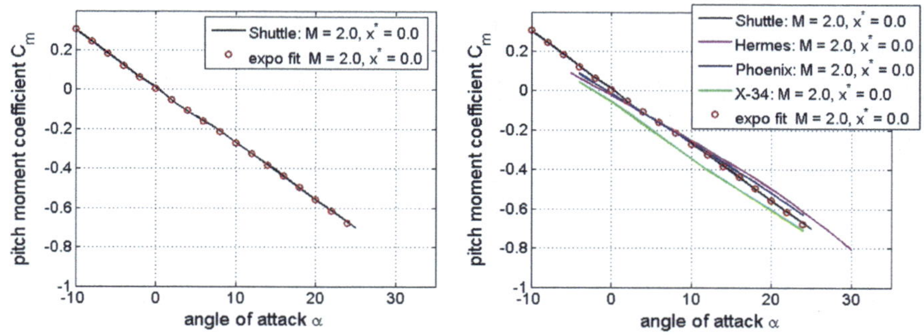

Fig. 7.40 Pitching moment coefficient C_m at $x_{ref} = x^* = 0$ for the supersonic Mach number $M_\infty = 2$, see Fig. 7.33. In the left figure the Shuttle graph is compared with the result of the fit equation 7.55 using $Cmsup$. The right figure contains additionally the data of Hermes, X-34 and Phoenix

is used to approximate the C_m function of the Shuttle Orbiter, Fig. 7.40 (left). In the right figure additionally the data for Hermes, X-34 and Phoenix are plotted. There obviously exists an acceptable agreement between the graphs of the Shuttle, Hermes and Phoenix. But besides the fact that the inclination of the X-34 curve is very similar to the Shuttle one, its position in the diagram is shifted by a negative ΔC_m for the X-34 vehicle.

7.7.4 Conclusions

We summarize the results of the above investigations as follows.

1. Oswatitsch's Mach number independence principle is impressively confirmed.

2. Re-entry vehicles

- The lift coefficients C_L at hypersonic Mach numbers for both WRV and LRV agree nearly perfectly: fully consistent similarity, Fig. 7.15.
- The lift coefficients C_L at supersonic Mach numbers for both WRV and LRV agree rather good: conditionally consistent similarity, Fig. 7.28.
- The drag coefficients C_D at hypersonic Mach numbers agree separately for WRV and LRV rather good: conditionally consistent similarity, Fig. 7.20.
- The drag coefficients C_D at supersonic Mach numbers agree for WRV[24] rather good: conditionally consistent similarity, Fig. 7.29.

The functions of the pitching moment coefficients of the various re-entry space vehicles depend on the defined reference positions x_{ref}. The definition of these quantities is more or less arbitrary. Therefore it makes no sense to compare these functions directly.

[24]Unfortunately there are no supersonic data available for the X-33 vehicle.

Table 7.3 Compilation of the characteristics regarding the general fit equation with its specific sets of coefficients. General fit equation for lift, drag and pitching moment coefficients, Eq. (7.55)

	M_∞	Coeff.	a_0	a_1	a_2	d_2	d_3	d_4
C_L	$M_\infty \gg 1$	$CLhyp$	1.0417	0.7546	−1.1033	1.15	2.30	2.15
C_D	$M_\infty \gg 1$ WRV	$CDhypW$	3.3460	−0.1922	−3.2662	1.00	0.72	2.50
	$M_\infty \gg 1$ LRV	$CDhypL$	2.7574	−0.0870	−2.6311	1.00	0.72	2.50
C_L	$M_\infty = 2$	$CLsup1$	4.7704	1.9992	−4.7902	0.50	0.72	1.75
	$M_\infty = 3$	$CLsup2$	8.9937	1.4698	−8.9747	0.50	0.72	2.10
	$M_\infty = 4$	$CLsup3$	7.4193	1.1703	−7.4669	0.50	0.72	1.90
C_D	$M_\infty = 2$	$CDsup1$	11.8871	−0.1037	−11.7552	0.50	0.72	2.00
	$M_\infty = 3$	$CDsup2$	11.7030	−0.1395	−11.5980	0.50	0.72	2.10
	$M_\infty = 4$	$CDsup3$	12.6368	−0.1453	−12.5434	0.50	0.72	2.20
C_m	$M_\infty \gg 1$	$Cmhyp$	−2.8074	−0.5107	2.8264	0.78	1.0	2.16
C_m	$M_\infty = 2$	$Cmsup$	−2, 6689	−1.6375	2.6745	0.78	1.0	2.16

In order to find anyhow a way to check the pitching moment coefficient functions, a transformation to a common reference point, for example to the nose of the configuration with $x^* = 0$ (Fig. 7.22), improves the situation. But often the origin of the x_{ref} values, and with that the origin of the coordinate system, is not clearly defined and therefore sometimes not in the nose. This is an ambiguous position and complicates the analysis of the pitching moment.

Nevertheless it was found that the pitching moment coefficients C_m for $x^* = 0$ in hypersonic flow agree to some extent, Fig. 7.26.

Further the slope of the pitching moment coefficients C_m for $x^* = 0$ in supersonic flow ($M_\infty = 2$) is very similar, but the graphs indicate a small spread, Fig. 7.33.

3. Hypersonic cruise vehicle

- The lift coefficient C_L at hypersonic Mach numbers agrees acceptably with the behavior of the re-entry vehicles: conditional similarity, Fig. 7.21(left).
- The lift coefficient C_L at supersonic Mach numbers agrees rather good with the behavior of the re-entry vehicles: conditionally consistent similarity, Fig. 7.28.
- The drag coefficient C_D at hypersonic Mach numbers has no conformity with the behavior of the re-entry vehicles: no similarity, Fig. 7.21(right).
- The drag coefficient C_D at supersonic Mach numbers has no conformity with the behavior of the re-entry vehicles: no similarity , Figs. 7.29 and 7.39.

In Table 7.3 we recapitulate the characteristics of the general fit equation with its specific sets of coefficients representing approximations of the lift, the drag and the pitching moment coefficients in supersonic and hypersonic Mach number regime of the space vehicles considered.

It seems that the analysis of the aerodynamic data of the space vehicles described above suggests a fully configurational independence of the lift coefficient C_L and a partly configurational independence of the drag coefficient C_D in hypersonic and supersonic Mach number regime.

For the pitching moment coefficient C_m with $x^* = 0$ the situation is somewhat more complicated, but also in this case some kind of independency can be observed.

The usefulness of the findings above consists in the potential, which scientists and engineers can exploit, who are involved in the development of future winged space planes and/or re-entry vehicles. If a minimum of data is available by wind tunnel tests or numerical simulations, most of the longitudinal aerodynamics in the supersonic and hypersonic Mach number regime can then simply be constructed with the related semi-empirical formulas presented above.

References

1. Zierep, J.: Theoretische Gasdynamik 1. Braun, Karlsruhe (1972)
2. Prandtl, L.: Vorlesungen Aerodynamik. Göttingen, Germany (1922)
3. Glauert, H.: Proc. R. Soc. A **118**, 113–119 (1928)
4. Göthert, B.: Jahrbuch der Luftfahrt **1**, 156–158 (1941)
5. Ackeret, J.: Habilitationsschrift, ETH Zürich (1928)
6. Busemann, A.: Jahrbuch der Wissenschaftlichen Gesellschaft für Luftfahrt, vol. 95 (1928)
7. Oswatitsch, K.: Advances in Applied Mechanics. Academic Press, vol. 6, 153–271 (1960)
8. Guderley, K.G.: Theory of Transonic Flow. Pergamon Press, New York (1962)
9. Keune, F., Burg, K.: Singularitätenverfahren der Strömungslehre. Braun Verlag, Karlsruhe (1975)
10. Tsien, H.S.: J. of Math. Phys. **25**, 247f (1946)
11. Oswatitsch, K.: ZAMP **2**, 249–264 (1951)
12. Zierep, J.: Theoretische Gasdynamik 2. Schallnahe und Hyperschallströmungen. Braun Verlag, Karlsruhe (1972)
13. Becker, E.: Gasdynamik. Teubner Verlag, Stuttgart (1965)
14. Krause, E.: Strömungslehre, Gasdynamik und aerodynamisches Laboratorium. Teubner Verlag, Stuttgart (2003)
15. Schlichting, H., Truckenbrodt, E.: Aerodynamik des Flugzeuges. Springer, Berlin (1967)
16. Zierep, J.: Ähnlichkeitsgesetze und Modellregeln der Strömungsmechanik. Braun Verlag, Karlsruhe (1972)
17. Oswatitsch, K.: Gas Dynamics. Academic Press, New York (1956)
18. von Kármán, Th.: International Congress for Applied Mechanics. Paper no. 6, Paris (1946)
19. Oswatitsch, K.: Royal Aircraft Establishment. Technical Note Aero 1902 (1947)
20. Keune, F.: Über die Kontinuitätsgleichung kompressibler Strömungen bei kleinen Störgeschwindigkeiten. Deutsche Forschungsanstalt für Luft- und Raumfahrt, DLR Mitteilungen, 70-13. Germany (1970)
21. Weiland, C.: A Contribution to the Computation of Transonic Supersonic Flows over Blunt Bodies. Computer & Fluids **9**, 143–162 (1981)
22. Weiland, C.: Die Berechnung stationärer reibungsfreier Strömungsfelder um stumpfe Körper im schallnahen Überschall mittels eines zeitabhängigen Differenzenverfahrens. In: Müller, U., Roesner, K.G., Schmidt, B. (eds.) Recent Developments in Theoretical and Experimental Fluid Mechanics, Springer, Heidelberg/New York, 267–276, (1979)
23. Liepmann, H.W., Roshko, A.: Elements of Gasdynamics. Wiley, New York (1957)
24. Hayes, W.D., Probstein, R.F.: Hypersonic Flow Theory. Academic Press, New York (1966)

25. Weiland, C.: Aerodynamic Data of Space Vehicles. Springer, Berlin (2014)
26. Weiland, C.: The aerodynamics of real space vehicles in the light of supersonic and hypersonic approximate theories. CEAS Space J. (2019). https://doi.org/10.1007/s12567-019-00264-w
27. Weiland, C.: Aerodynamic similarity of winged and lifting space vehicles. AIAA J. Aircraft. https://doi.org/10.2514/1.C035823 (2020). Accessed 30 Nov 2020
28. Hirschel, E.H.: Basics of Aerothermodynamics, 2nd edn. Springer, Berlin (2015)
29. Hirschel, E.H., Weiland, C.: Selected Aerothermodynamic Design Problems of Hypersonic Flight Vehicles. Springer, Berlin and AIAA (Progress in Astronautics and Aeronautics), Reston USA (2009)

Chapter 8
Model Test Entity

When humans have built large technical devices like bridges, ships, buildings, embankment dams, airplanes, space planes, reactors for chemical purposes, hydraulic engines (pumps, compressors, turbines) etc. there was a need to fabricate geometrically similar models of these devices in order to investigate and test their physical behavior in ground based test facilities. The main question from the physical point of view in this regard is under what conditions are the experimental test results, received with geometrically similar models in test facilities, transferable to the situation of the original devices? We define the three similarity rules, which must be kept for a suitable testing of the physics of originals in test facilities, [1, 2].

8.1 The Geometrical Similarity

Every arbitrary vector (e.g. $\mathbf{r} = (x, y, z)^T$) of a length of an original (1) is scaled by a transformation factor $\alpha_1 > 0$ to receive a model (2), which is then geometrically similar

$$\mathbf{r}_1 = \alpha_1 \mathbf{r}_2 \quad \text{or} \quad \begin{pmatrix} x_1 \\ y_1 \\ z_1 \end{pmatrix} = \alpha_1 \begin{pmatrix} x_2 \\ y_2 \\ z_2 \end{pmatrix} .$$

8.2 The Kinematic Similarity

Related times in an original system are transformed to the related times in a model system by the factor $\alpha_2 > 0$

C. Weiland, *Mechanics of Flow Similarities*,
https://doi.org/10.1007/978-3-030-42930-0_8

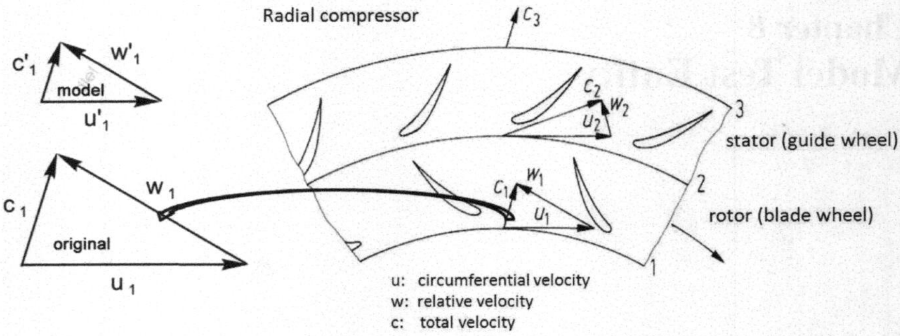

Fig. 8.1 Transformation of velocity vectors: Triangles of velocity vectors in a fluid flow engine (compressor). Model velocities written with a prime. Sketch taken from [3]

$$t_1 = \alpha_2 t_2 \, .$$

This means that velocities in both systems are transformed by

$$u_1 = \beta_1 u_2 \quad \text{with} \quad \beta_1 = \frac{\alpha_1}{\alpha_2} \quad \text{or} \quad \beta_1 \alpha_2 \alpha_1^{-1} = 1 \, .$$

This guarantees that, for example, the streamline pattern of a flow is similar in both systems. For demonstration Fig. 8.1 shows the flow situation of the triangles of the velocity vectors of a fluid flow engine (compressor), which are similar for the original and the model, [3].

8.3 The Dynamic Similarity

A further requirement for the similarity between an original and a model system affects the acting forces.[1] A model system is then similar to an original system, if the forces at related locations and at related times are similar. Let the masses be transformed with $\alpha_3 > 0$ by

$$m_1 = \alpha_3 m_2 \, ,$$

and with the acceleration

$$a_1 = \frac{\alpha_1}{\alpha_2^2} a_2 \, ,$$

[1]These are for example inertial, gravitational, friction and pressure forces.

one obtains for the forces

$$m_1 a_1 = \mathbf{F}_1 = \alpha_3 \frac{\alpha_1}{\alpha_2^2} m_2 a_2 = \beta_2 \mathbf{F}_2 \, .$$

The dynamic model law then reads, if the choice of the set of basic dimensions is left open (see Sect. 2.2)

$$\beta_2 \, \alpha_2^2 \, \alpha_1^{-1} \, \alpha_3^{-1} = 1 \, .$$

Considering a specific flow field, where various forces are acting, like inertial, friction, gravitational and pressure forces, the dynamic similarity is present for all these forces. In Sect. 3.1 we found that the inertial force is proportional to $\rho u^2/l$ and the friction force to $\mu u/l^2$. This results for the inertial force in the transformation rule

$$\left(\frac{\rho u^2}{l}\right)_1 = \frac{1}{\alpha_1}\left(\frac{\alpha_1}{\alpha_2}\right)^2 \frac{\alpha_3}{\alpha_1^3}\left(\frac{\rho u^2}{l}\right)_2 = \frac{\alpha_3}{\alpha_1^2\alpha_2^2}\left(\frac{\rho u^2}{l}\right)_2 , \tag{8.1}$$

which must also be valid for the friction force. Therefore we obtain

$$\left(\frac{\mu u}{l^2}\right)_1 = \frac{\alpha_3}{\alpha_1^2\alpha_2^2}\left(\frac{\mu u}{l^2}\right)_2 \quad \text{or} \quad \frac{\mu_1}{\mu_2} = \frac{u_2}{u_1}\frac{l_1^2}{l_2^2}\frac{\alpha_3}{\alpha_1^2\alpha_2^2} = \frac{\alpha_3}{\alpha_1\alpha_2} ,$$

and with[2]

$$\mu_1 = \frac{\rho_1}{\rho_2}\frac{l_1}{l_2}\frac{u_1}{u_2}\,\mu_2 ,$$

it finally follows the Reynolds number

$$\frac{\rho_1 u_1 l_1}{\mu_1} = \frac{\rho_2 u_2 l_2}{\mu_2} = Re \, .$$

Of course Eq. (8.1) is also valid for the gravitational force, which is proportional to $\rho\,\mathbf{g}$ and we obtain with

$$(\rho g)_1 = \frac{\alpha_3}{\alpha_1^2\alpha_2^2}(\rho g)_2 ,$$

for the ratio of inertial force to gravitational force

$$\begin{aligned} \frac{\rho_1}{\rho_2}\frac{g_1}{g_2} = \frac{\alpha_3}{\alpha_1^2\alpha_2^2} &= \frac{\alpha_3}{\alpha_1^3}\cdot\frac{1}{\alpha_1}\cdot\frac{\alpha_1^2}{\alpha_2^2} \\ &= \frac{\rho_1}{\rho_2}\cdot\frac{l_2}{l_1}\cdot\frac{u_1^2}{u_2^2} \end{aligned}$$

[2]The relation for the density is given by $\rho_1/\rho_2 = \alpha_3/\alpha_1^3$.

which is the Froude number

$$\frac{u_1^2}{l_1 g_1} = \frac{u_2^2}{l_2 g_2} = Fr\ .$$

In a model test the aim is to duplicate all the relevant similarity parameters. Supposed that the flow is governed by the Reynolds number and the Froude number. Example are the test of a ship design in a water tunnel, or the flow over a barrage. Then the reduction of the typical length of a model l_2 must be compensated by an increase of the velocity u_2 and/or a reduction of the kinematical viscosity $\nu_2 = \mu_2/\rho_2$ in order to satisfy the Reynolds similarity. However for the Froude similarity the reduction of l_2 can only be compensated by a decrease of the velocity u_2. It seems to be impossible to solve this dilemma. Therefore the experimentalist or test engineer has to decide, which similarity is more significant to be duplicated for the flow problem considered.

In principle all the dimensionless numbers generated by the dimensional analysis, Chap. 2, can be used to provide tests with similar physical proportions, i.e. model tests, independently of the question whether a model system exists and how it could look like. This includes also those which are connected with the evaluation of the energy equation (all sorts of heat transfer problems), Sect. 3.2. Nevertheless model tests are mainly conducted in the fluid dynamic discipline with its large number of wind tunnels covering wide ranges of the Mach number and the Reynolds number and furthermore in water tunnels with the Froude number as governing quantity. It should be mentioned that beyond the above statements there are additional parameters (with dimensions!), from which experience has revealed that they describe similar physical processes.[3]

8.4 Demonstration of the Significance of Model Testing

The aerodynamics of conventional (civil or military) airplanes is governed by the Reynolds number Re and the Mach number M. In order to consider similar flow fields, these both numbers should be duplicated in wind tunnel tests. Whereas the Mach number is often held in wind tunnel investigations, the reproduction of the Reynolds number for large aircrafts is rather difficult. Large wind tunnels, where big models can be tested (like the Large Low Speed Facility LLF of the German/Dutch organisation DNW) ease this problem. For transonic Mach numbers the European Transonic Wind tunnel ETW, a hugh facility, enables to reproduce both the Mach number and the Reynolds number (up to $Re \approx 8.5 \cdot 10^7$ in its test chamber), which is sufficient

[3]Good examples are the parameter $\rho \cdot L$, which comes into play for hypersonic flows with non-equilibrium real gas effects (binary scaling parameter), and also the rarefaction parameter $\bar{V} = M/\sqrt{Re \cdot L}$, characterising the separation length of concave flow situations (e.g.: deflected aerodynamic control surfaces) in high enthalpy flows, [4, 5].

even for the Airbus A380, see Sect. 6.5. Nevertheless most of the conventional wind tunnels cannot duplicate the Reynolds number.

But there are other effects which derogate wind tunnel tests. The test section has its influence due to its finite extent compared to the flight in the atmosphere. The free-stream must be uniform. Wall interferences of several kinds, model and model-support presence as well as possible model deformations have also to be taken into account. Two special phenomena are not covered by similarity parameters:

- the free-stream turbulence,
- the roughness of the model surface.

Both have essential influence on the laminar-turbulent transition process, which is known as receptivity problem, [6, 7]. If the Reynolds number is not large enough, boundary layer tripping must be applied in order to achieve turbulent flow. Moreover, if the boundary layer is turbulent, the surface roughness must be sub-critical in view of the shear stress and the heat transfer at the wall. These both quantities are very strongly enlarged and falsified by a super-critical surface roughness.

High supersonic and hypersonic tests in ground facilities embody other problems. The free-stream in the wind tunnel nozzle may be in a "frozen" thermodynamic state. High temperature real gas effects as well as thermal and chemical rate processes may occur and can be prescribed by the two Damköhler numbers. The wall temperature, more general the thermal state of the surface, influences the boundary layer, laminar-turbulent transition, wall shear stress, flow separation, surface catalytic recombination, all together called "thermal surface effects", [8]. Further in reality surface radiation cooling is present, or even active cooling for instance by ablation processes. This cannot be simulated in ground simulation facilities.

Typically these problems are present in re-entry space flight. We encounter there a wide range of Mach numbers ($0 < M_\infty \lesssim 30$) and Reynolds numbers ($0 < Re \lesssim 5 \cdot 10^7$) as well as various states of the thermodynamics, as described above, of the air to be passed. To demonstrate the problem we consider the X-38 re-entry flight along the trajectory shown in Fig. 8.2, [9].

In order to state what can be expected from model tests in hypersonic test facilities the conditions in the three "cold"[4] hypersonic wind tunnels H2K and RWG of the DLR[5] as well as S4MA of ONERA[6] and the two high enthalpy facilities HEG (DLR) and F4 (ONERA) are considered and compared with the flight data of the X-38 re-entry trajectory, [9]. In Table 8.1 the test conditions are listed. The Reynolds numbers were determined with the typical model lengths appropriate in the various test facilities.

---

[4]Cold means that the freestream total temperature or enthalpy is much lower than the one the vehicle encounters during real flight.

[5]German Aerospace Establishment.

[6]Office National D'Etudes et de Recherches Aerospatiales, France.

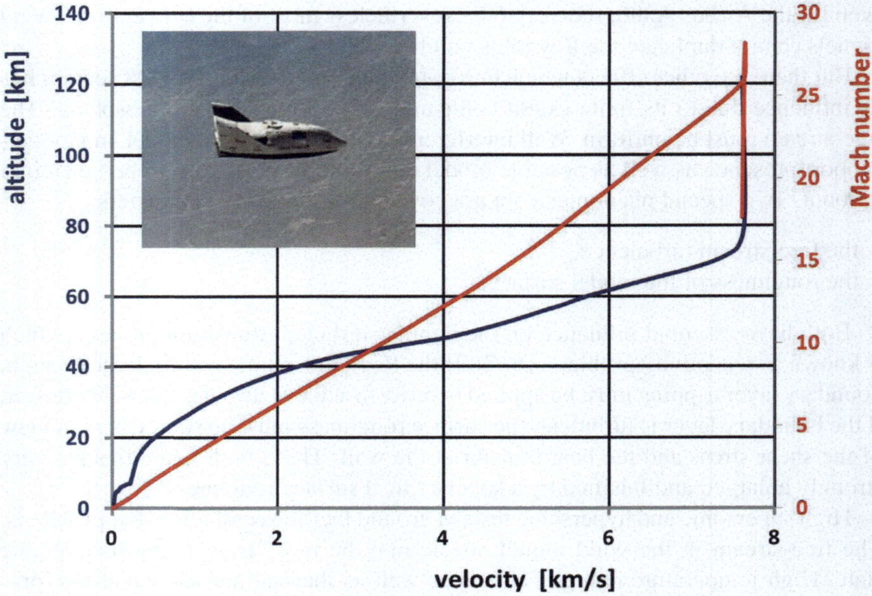

Fig. 8.2 X-38 re-entry trajectory, [9]

Table 8.1 Test conditions in various hypersonic wind tunnels, [4, 10–12]. Assumed model lengthes: H2K ⇒ 0.50 m, RWG ⇒ 0.25 m, S4MA ⇒ 0.42 m, HEG ⇒ 0.40 m, F4 ⇒ 0.40 m

Facility	Re	M_∞	v_∞ [m/s]	T_∞ [K]	p_∞ [Pa]	ρ_∞ [kg/m^3]
H2K (1)	$0.310 \cdot 10^6$	8.70	1475	71.3	41.5	$2.0 \cdot 10^{-3}$
H2K (2)	$0.600 \cdot 10^6$	8.70	1305	55.8	53.3	$3.3 \cdot 10^{-3}$
RWG (1)	$0.875 \cdot 10^6$	6.83	1045	58.1	225.8	$13 \cdot 10^{-3}$
RWG (2)	$1.875 \cdot 10^6$	6.83	1045	58.1	451.7	$26 \cdot 10^{-3}$
RWG (3)	$3.750 \cdot 10^6$	6.83	1045	58.1	903.3	$53 \cdot 10^{-3}$
S4MA (1)	$0.833 \cdot 10^6$	9.77	1417	52.2	71.2	–
S4MA (2)	$2.617 \cdot 10^6$	9.95	1418	50.5	207.1	–
HEG (I)	$0.108 \cdot 10^6$	9.70	5919	790	430.0	$1.64 \cdot 10^{-3}$
HEG (III)	$0.191 \cdot 10^6$	9.98	4813	553	470.0	$2.93 \cdot 10^{-3}$
F4 (case 1)	$0.194 \cdot 10^6$	14.00	3934	178	–	$1.56 \cdot 10^{-3}$
F4 (case 2)	$0.070 \cdot 10^6$	9.90	4382	471	–	$1.03 \cdot 10^{-3}$
F4 (case 5)	$0.022 \cdot 10^6$	7.20	6055	1870	–	$0.55 \cdot 10^{-3}$

The main objective of wind tunnel investigations in industrial development projects concerning flight vehicles is the measurement of all the coefficients, which describe the steady and unsteady aerodynamic behavior of the flight or space vehicle, [13]. This set of coefficients should guarantee a reliable and controlled flight operation. Of course when the Reynolds number can not be completely duplicated, the data are potentially not fully consistent with the real flight data whereby an uncertainty is generated. The usual way to overcome this problem persists in a verification procedure of these data, for example by flight tests. Another possibility of verification is the application of numerical methods (CFD), which provides the design engineer (or scientist) with the solution of complete flowfields, from which all the aerodynamic coefficients can be extracted. Further the numerical methods offer the potential, if their results are verified with the test conditions of ground based facilities, to compute the flow fields with the realistic free flight conditions (In this sense the numerical methods act as transfer models from test to free flight.).

The argumentation above is also true for flows where the thermodynamic behavior plays a major role in the formation of all heat transfer aspects, in particular the thermal surface effects, see Chap. 10 of [8], and the state of hot gases (equilibrium and non-equilibrium real gas effects including the interactions with the vehicle walls (catalycity)). This is definitely the case for hypersonic and high enthalpy flows.

In so far the flows in the hypersonic regime (including the high enthalpy range) can be characterized mainly by the dimensionless numbers Re, M, $(T_w - T_\infty)/T_\infty$, the similarity parameter $\rho \cdot L$ as well as the flight speed $\mathbf{v}$ or the total enthalpy h_0, respectively. As long as the flight velocity $\mathbf{v}$ is smaller than approximately 1500 m/s flow similarity is govern by Re, M, and $(T_w - T_\infty)/T_\infty$. For larger velocities and higher altitudes the importance of the Mach number is getting less. The dimensionless parameter $(T_w - T_\infty)/T_\infty$ is also a measure for the influence a hot surface[7] may achieve on the state of the boundary layer, which then results in a decrease of the viscous drag of the space vehicle, [8].

When we have a look on the X-38 Reynolds-Mach diagram, Fig. 8.3, where the wind tunnel conditions of H2K, RWG and S4MA are sketched in, we notice that only one condition of RWG and one of S4MA for the Reynolds number are in the vicinity of the flight trajectory.

The experience has revealed that for increasing flight velocities (increase of total enthalpy) the role of the Mach number, characterizing the flow field, diminishes and is replaced by the velocity (or the total enthalpy). Therefore flow similarity can properly be achieved in this flight regime,—beyond an altitude of $\approx$ 50 km -, by tests in high enthalpy facilities (e.g. HEG and F4), Figs. 8.2 and 8.4. The test conditions in HEG and F4 reflect the high velocity (or high enthalpy) capacity of these facilities, Table 8.1.

As already mentioned to attain similarity with points of realistic flight trajectories the investigations in hypersonic wind tunnels like H2K, RWG and S4MA should duplicate the dimensionless numbers Re, M and $(T_w - T_\infty)/T_\infty$. All test facilities considered here use models, where the vehicle walls have temperatures of the

[7]Generation of hot surfaces during re-entry flight of space vehicles.

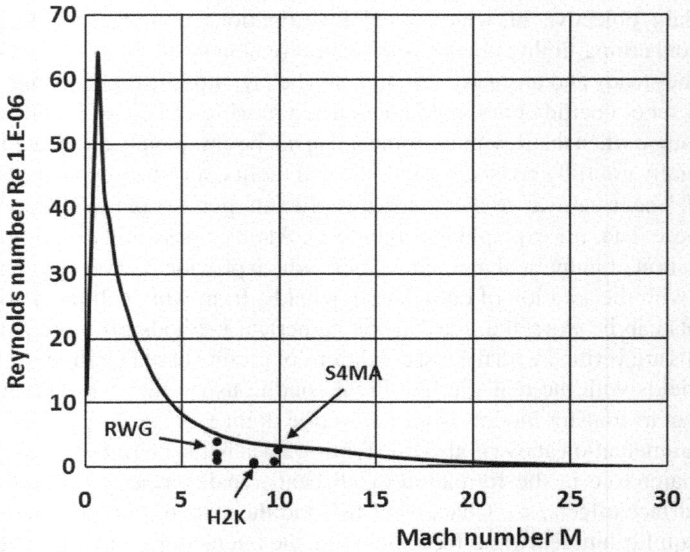

Fig. 8.3 Reynolds number as function of the Mach number taken from the X-38 re-entry trajectory, [9]. Included are the test conditions of the cold hypersonic wind tunnels RWG, H2K, S4MA, [10, 11], see Table 8.1

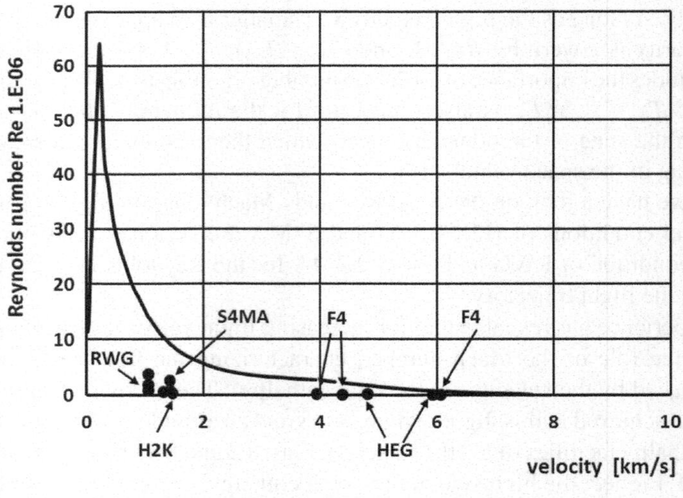

Fig. 8.4 Reynolds number as function of the velocity taken from the X-38 re-entry trajectory, [9]. Included are the test conditions of the three cold hypersonic wind tunnels RWG, H2K, S4MA, and the two high enthalpy facilities HEG and F4, [4, 10–12], see Table 8.1

Table 8.2 Test facility conditions reflected on related X-38 flight trajectory points

	Re	M_∞	v_∞ [m/s]	T_∞ [K]	H [km]
RWG (3)	$3.750 \cdot 10^6$	6.83	1045	58.1	–
$X38^{RWG}_{Mach}$	$4.831 \cdot 10^6$	**6.98**	2226	251.0	40
$X38^{RWG}_{Re}$	**3.751 10^6**	8.51	2752	257.0	43
S4MA (2)	$2.616 \cdot 10^6$	9.95	1418	50.5	–
$X38^{S4MA}_{Mach}$	$3.220 \cdot 10^6$	**10.01**	3263	261.0	45
$X38^{S4MA}_{Re}$	**2.607 10^6**	12.49	4084	262.0	48
HEG (III)	$0.191 \cdot 10^6$	9.98	4813	553	–
$X38^{HEG}_{Mach}$	$3.220 \cdot 10^6$	**10.01**	3263	261.0	45
$X38^{HEG}_{Re}$	**0.214 10^6**	25.00	7449	219.0	72
$X38^{HEG}_{velo}$	$1.711 \cdot 10^6$	14.99	**4851**	256.0	53
F4 (2)	$0.070 \cdot 10^6$	9.90	4382	471	–
$X38^{F4}_{Mach}$	$3.220 \cdot 10^6$	**10.01**	3263	261.0	45
$X38^{F4}_{Re}$	**0.089 10^6**	26.00	7623	213.0	78
$X38^{F4}_{velo}$	$2.188 \cdot 10^6$	13.51	**4417**	262.0	50

ambience (typically $\approx$ 293 K). But in real hypersonic flight the walls of flight vehicles heat up to more than 1000 K, [8, 14], which obviously cannot be duplicated in the test facilities. Therefore we do not observe further the parameter $(T_w - T_\infty)/T_\infty$. As Table 8.1 has indicated to duplicate Re and M at the same time is nearly impossible.

To highlight the problem, we have extracted from the X-38 trajectory the points where first the trajectory Mach number (M_{traj}) is next to the test Mach number (M_{test}) (alternatively the velocity **v**) and second the trajectory Reynolds number (Re_{traj}) is nearest to the test Reynold number (Re_{test}), see Table 8.2.

- RWG, Mach number duplication: $Re_{traj} \approx 28\%$ larger than Re_{test},
- RWG, Reynolds number duplication: $M_{traj} \approx 24\%$ larger than M_{test},
- S4MA, Mach number duplication: $Re_{traj} \approx 23\%$ larger than Re_{test},
- S4MA, Reynolds number duplication: $M_{traj} \approx 25\%$ larger than M_{test},
- HEG, Reynolds number duplication: $v_{traj} \approx 55\%$ larger than v_{test},
- HEG, velocity duplication: $Re_{traj} \approx 795\%$ larger than Re_{test},
- F4, Reynolds number duplication: $v_{traj} \approx 74\%$ larger than v_{test},
- F4, velocity duplication: $Re_{traj} \approx 3030\%$ larger than Re_{test}.

The example above describes the problems of model tests on the basis of flow similarity can have, but it says nothing about how much the incompleteness of the modelling influences the accuracy of the measurement of the aerodynamic (or aerothermodynamic) data. To check this is always the task of a verification process.

When the flow behind a bow shock is in a non-equilibrium thermodynamic state it was observed that the dimensional quantity $\rho \cdot L$ can act as similarity parameter.

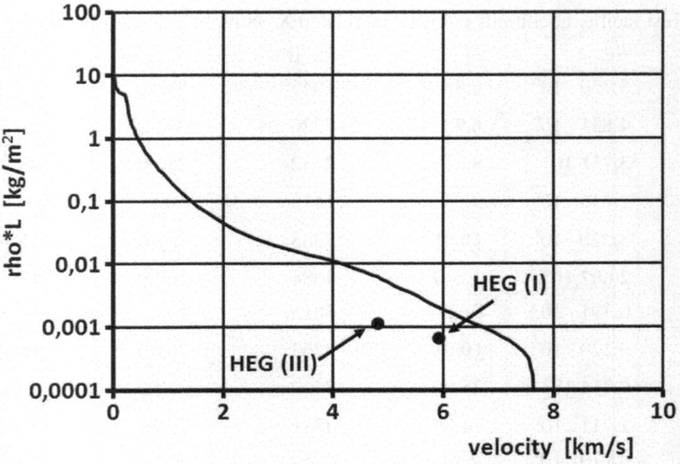

Fig. 8.5 $\rho * L$ as function of the velocity taken from the X-38 re-entry trajectory, [9]. Included are the test conditions of the high enthalpy facility HEG, [4]

In Fig. 8.5 the test conditions of HEG are plotted together with the X-38 trajectory data, which reveals that a complete duplication in HEG can not be achieved. Besides others the $\rho \cdot L$ similarity is important for the estimation of wall catalytic effects.

References

1. Braun, J.: Strömungsmaschinen als Kraft- und Arbeitsmaschinen, Teil 1 Hydraulische Maschinen. Books on Demand, Norderstedt, Germany (2015)
2. Görtler, H.: Dimensionsanalyse. Theorie der physikalischen Dimensionen und Anwendungen. Springer, Berlin (1975)
3. Beitz, W., Küttner, K.-H. (eds.): Dubbel: Taschenbuch für den Maschinenbau, 17th edn. Springer, Berlin (1990)
4. Krek, R.M., Eitelberg, G., Kastell, D.: Hyperboloid Flare Experiments in the HEG Facility. Deutsche Forschungsanstalt für Luft- und Raumfahrt, DLR-IB 223-95 A 43 (1995)
5. Devezeaux, D., Tribot, J.P.: Synthesis of contribution to the hyperboloid-flare F4 test case TC1.C within the Manned Space Transportation Program Workshop 1996. ONERA, Technical Report N° RT 92/6121 SY (1997)
6. Hirschel, E.H.: Three-Dimensional Attached Viscous Flow. Springer, Berlin (2014)
7. Schlichting, H., Truckenbrodt, E.: Aerodynamik des Flugzeuges, vol. 1. Springer, Berlin (1967)
8. Hirschel, E.H.: Basics of Aerothermodynamics, 2nd edn. Springer, Berlin (2015)
9. Weiland, C.: Computational Space Flight Mechanics. Springer, Berlin (2010)
10. Brenner, G., Hannemann, K., Kordulla, W.: The Hyperboloid-Flare Experiment - Summary of numerical results. DLR Internal Report IB 221-93 C 28 (1993)
11. Bousquet, J.M., Faubert, A.: Synthesis of hyperboloid-flare results in R5Ch, R3Ch and S4MA wind tunnel conditions. ONERA, Rapport Technique n° RT 63/6122 AY, HT-TN-E-1-314-ONER (1995)

12. Sagnier, P.: Synthesis of hyperboloid-flare experiments in ONERA F4, S4MA, R3Ch, R5Ch wind tunnels and DLR H2K, RWG wind tunnels. ONERA, Rapport Technique n° RT 72/6121 SY (1995)
13. Weiland, C.: Aerodynamic Data of Space Vehicles. Springer, Berlin (2014)
14. Hirschel, E.H., Weiland, C.: Selected Aerothermodynamic Design Problems of Hypersonic Flight Vehicles. Springer, Berlin and AIAA (Progress in Astronautics and Aeronautics), Reston USA (2009)

Appendix
Symbols

Only the important symbols are listed. In general the page number is indicated, where a symbol appears first or is defined. Dimensions are given in terms of the SI basic units.

A.1 Latin Letters

		page	
$\mathbf{A}$	matrix (Π-theorem)	6	[−]
A	outflow area	10	[m^2]
A	axial force	147	[N]
Ar	Archimedes number	117	[−]
a	speed of sound	132	[m/s]
a	coefficient of similarity transformation	135	[−]
a	scaling factor	152	[−]
a_1, a_2	accelerations	177	[m/s^2]
a_0, a_1, a_2	coefficients of semi-empirical formula	165	[−]
Bi	Biot number	118	[−]
Bm	Bingham number	118	[−]
Bo	Bodenstein number	119	[−]
Br	Brinkman number	119	[−]
$\mathbf{b}$	vector (Π-theorem)	6	[−]
b_n	vector elements (Π-theorem)	6	[−]
b	acceleration	8	[m/s^2]
C	coefficient of similarity transformation	135	[−]
C_A	axial force coefficient	156	[−]
C_D	drag coefficient	147	[−]
C_i	aerodynamic coefficients	165	[−]

C. Weiland, *Mechanics of Flow Similarities*,
https://doi.org/10.1007/978-3-030-42930-0

C_L	lift coefficient	147	[−]
C_m	pitching moment coefficient	156	[−]
C_N	normal force coefficient	156	[−]
c_p	spec. heat at constant pressure	15	$[m^2/s^2K]$
c_p	pressure coefficient	134	[−]
c_v	spec. heat at constant volume	15	$[m^2/s^2K]$
D	pipe diameter	13	[m]
D	drag force	14	$[N]$
D_{AB}	binary mass diffusivity	35	$[m^2/s]$
$DAM1$	1. Damköhler number	43	[−]
$DAM2$	2. Damköhler number	41	[−]
d	shock stand-off distance	75	[m]
d	maximal thickness of a thin body	134	[m]
d_2, d_3, d_4	coefficients of semi-empirical formula	165	[−]
E_{kin}	kinetic energy	64	$[kgm^2/s^2]$
Ec	Eckert number	41	[−]
Eu	Euler number	40	[−]
e	total energy per unit mass	25	$[m^2/s^2]$
e_j	heat fluxes per unit volume	27	$[kg/s^3m]$
e_v	internal energy per unit mass	25	$[m^2/s^2]$
F	surface	30	$[m^2]$
F	$F(x, y, z)$ contour function	132	[−]
F_j	forces	24	$[N]$
Fo	Fourier number	42	[−]
Fr	Froude number	40	[−]
$d\mathbf{F}$	surface element	30	$[m^2]$
$\mathbf{f}_e$	vector of external forces per unit mass	30	$[m/s^2]$
f	$f(x)$ x-dependent part of contour function	133	[−]
f	force	20	$[N]$
f	frequency of wave	55	$[1/s]$
f, f_i	functions (Π-theorem)	7	[−]
Ga	Galilei number	120	[−]
Gr	Grashof number	40	[−]
g, g_i	functions (Π-theorem)	11	[−]
g	gravitational acceleration	9	$[m/s^2]$
g_x	x-component of gravitational acceleration	33	$[m/s^2]$
H	altitude in atmosphere	114	[m]
h	altitude	10	[m]
J	momentum flux	63	$[kg/s^2m]$
K	Tsien parameter	146	[−]
Kn	Knudsen number	43	[−]
K_i	physical variable (Π-theorem)	5	[−]
k_i	coefficients (Π-theorem)	8	[−]
k	roughness	13	[m]

k	thermal diffusivity	18	[m^2/s]
k	Boltzmann constant ($1.38 \cdot 10^{-23}$)	109	[Nm/K]
k_t	heat transfer eddy viscosity	91	[−]
L	lift force	152	[N]
La	Laplace number	120	[−]
Le	Lewis number	42	[−]
L_i	physical variable (Π-theorem)	6	[−]
l	length	8	[m]
l	l_{ref} reference length	134	[m]
M	Mach number	43	[−]
m_a	mass	8	[kg]
m_1, m_2	masses	177	[kg]
N	normal force	147	[N]
Nu	Nusselt number	41	[−]
$\mathbf{n}$	surface normal	132	[−]
Oh	Ohnesorge number	120	[−]
Pe	Peclet number	41	[−]
Pr	Prandtl number	42	[−]
p	number of dimensionless parameters (Π-theorem)	7	[−]
p	pressure	23	[N/m^2]
Δp	pressure difference	8	[N/m^2]
Q	amount of heat per volume	73	[kg/s^2m]
$\dot{Q}$	heat flux per unit volume	27	[kg/s^3m]
q	dynamic pressure	53	[kg/s^2m]
q_H	heat flux per unit volume	25	[kg/s^3m]
q_w	heat flux into the wall per surface element	77	[kg/s^3]
R	specific gas constant	113	[m^2/s^2K]
$\mathcal{R}$	maximum thickness radius	139	[m]
Ra	Rayleigh number	44	[−]
Re	Reynolds number	40	[−]
Ri	Richardson number	121	[−]
Ro	Roshko number	121	[−]
$\mathbf{r}$	position vector	130	[−]
r	rank of matrix (Π-theorem)	7	[−]
Sc	Schmidt number	42	[−]
Sh	Sherwood number	121	[−]
St	Stanton number	41	[−]
Ste	Stefan number	41	[−]
Sto	Stokes number	44	[−]
Str	Strouhal number	42	[−]
S_{ref}	reference area	152	[m^2]
S^*_{ref}	homogenized reference area	152	[m^2]
s	entropy	131	[m^2/s^2K]

T	temperature	15	$[K]$
T	period of wave	55	$[s]$
t	time	8	$[s]$
u, v, w	components of velocity vector	23	[m/s]
V	volume	30	$[m^3]$
$d\mathbf{V}$	volume element	30	$[m^3]$
$\overline{V}$	rarefaction parameter	178	$[m^{-0.5}]$
v	velocity	8	[m/s]
We	Weber number	40	[−]
x_{ref}, z_{ref}	individual moment reference coordinates	156	[m]
x^*, z^*	moment reference coordinates in body nose	156	[m]
x, y, z	Cartesian coordinates	23	[m]
x, r, θ	cylindrical coordinates	138	[−]
$\overline{\overline{I}}$	unit tensor	30	[−]

A.2 Greek Letters

The Greek alphabet reads

$\alpha, \beta, \gamma, \delta, \epsilon, \zeta, \eta, \vartheta, \iota, \kappa, \lambda, \mu, \nu, \xi, o, \pi, \phi, \rho, \sigma, \tau, \varphi, \chi, \upsilon, \psi, \omega.$

		page	
α	heat transfer coefficient	15	[N/msK]
α	degree of dissociation	76	[−]
α	angle of attack	146	[°]
α_{mn}	matrix elements (Π-theorem)	6	[°]
α_1	coefficient of geometrical transformation	130	[−]
$\alpha_1, \alpha_2, \alpha_3$	similarity coefficients	177	[−]
β	volumetric coefficient of thermal expansion	59	$[1/K]$
β	Prandtl-Glauert factor	133	[−]
β	yaw angle	147	[°]
β_1, β_2	coefficients of geometrical transformation	130	[−]
β_1, β_2	similarity coefficients	177	[−]
γ	ratio of specific heats	17	[−]
δ	boundary layer thickness	19	[m]
δ_T	thermal boundary layer thickness	47	[m]
ϑ	flow angle, turning angle	146	[°]
λ	thermal conductivity	15	[N/sK]
λ	wave length	55	[m]
λ	free mean path	109	[m]
μ	dynamic viscosity	8	$[Ns/m^2]$
ν	kinematic viscosity	18	$[m^2/s]$

ν_t	momentum eddy viscosity	91	[−]
ϕ	potential function of flow field	131	[m^2/s]
φ	small perturbation potential	133	[m^2/s]
Π_i	dimensionless parameters (Π-theorem)	7	[−]
ρ	density	8	[kg/m^3]
σ	surface tension	25	[N/m]
σ	diameter of a molecule of air ($3.7 \cdot 10^{-10}$)	109	[m]
$\bar{\sigma}$	Stefan–Boltzmann constant ($5.688 \cdot 10^{-8}$)	27	[W/m^2K^4]
$\bar{\bar{\tau}}$	viscous stress tensor	30	[−]
τ	thickness ratio	148	[−]
τ_{ij}	components of viscous stress tensor	23	[N/m^2]
χ	transonic similarity parameter	145	[m]
ω	mass fraction of species	35	[−]

Subject Index

C. Weiland, *Mechanics of Flow Similarities*,
https://doi.org/10.1007/978-3-030-42930-0

Author Index

C. Weiland, *Mechanics of Flow Similarities*,
https://doi.org/10.1007/978-3-030-42930-0

The manufacturer's authorised representative in the EU is Springer Nature Customer Service Centre GmbH, Europaplatz 3, 69115 Heidelberg, Germany. If you have any concerns regarding our products, please contact ProductSafety@springernature.com

Printed and bound by CPI Group (UK) Ltd, Croydon, CR0 4YY
05/10/2026
02228126-0001